Computers in
Botanical Collections

Computers in Botanical Collections

Edited by

J. P. M. Brenan

Royal Botanical Gardens
Kew, England

R. Ross

British Museum (Natural History)
London, England

and

J. T. Williams

University of Birmingham
Birmingham, England

Published in coordination with NATO Scientific
Affairs Division by

PLENUM PRESS • LONDON AND NEW YORK

Proceedings of an international conference, sponsored by NATO, on *The Use of Electronic Data Processing: Major Plant Taxonomic Collections*, held at Kew in October 1973

Library of Congress Catalog Card Number 75-9386
ISBN-13: 978-1-4684-2159-0   e-ISBN-13: 978-1-4684-2157-6
DOI: 10.1007/978-1-4684-2157-6

PREFACE

This volume records the proceedings of, and the papers read at, an international conference to consider the use of electronic data processing methods in the major taxonomic plant collections of Europe, primarily herbaria but also living collections. This conference took place at the Royal Botanic Gardens, Kew, from 3rd to 6th October, 1973. It was attended by some 90 delegates, observers and speakers, mainly from a wide range of the major European herbaria, but also from other interested institutions.

The problem to be discussed was a big one. Taxonomic collections of the sorts mentioned above constitute the main centres for the scientific documentation of the flora of the world. With the extinction of so many species more or less imminently threatened, and with the modification or disappearance of so many vegetation types through the activities of man, the information contained in these collections grows in importance. Their aggregate size, in Europe, has been estimated at between 50 and 100 millions, and these are annually augmented at a rapid rate. Each specimen or living plant comprises a source of evidence and information represented both by the specimen itself and the associated information provided by the collector and subsequent investigators - identifications, field notes, records of subsequent research, etc. This associated information is known as the label data.

Research on the taxonomy and geographical distribution of plants usually requires data from specimens stored in a number of institutions in different countries. At present it is difficult or impossible to survey this information completely.

It is now feasible, with the help of large computers, to compile comprehensive catalogues of much of this information, and to make it more widely available in the form of tapes or print-out. There is, however, the danger that, by independent action, institutions may adopt incompatible systems which can result in the stored information becoming unavailable internationally.

Considerable progress has been made in the United States, but in Europe, where perhaps the majority of major herbaria are located, progress in the application of E.D.P. methods in herbaria has been hitherto slight. Furthermore, most schemes have been concerned with relatively small samples - up to 50, 000 to 100, 000 items. Major schemes, even if desirable, have been too demanding in time, money and staff. The problem remains, however, and increases rather than diminishes, but funding through the international agencies may represent a future solution.

It was to study the scope of E.D.P methods in the major collections and to give an impetus to their application in Europe that this conference was held. It should be made clear that the objective was to discuss problems connected with the storage and retrieval of label data, and not the handling of taxonomic data for such purposes as numerical taxonomy, mechanised key construction, automated identification, etc.

The conference was made possible by financial support from the Eco-Sciences Panel of NATO, to whom our thanks are due. Gratitude is also owed to Dr. Andreas Rannestad, Secretary of the Eco-Sciences Panel, who was an unfailing source of help and advice during the organisation of the conference and who represented NATO at the meeting itself.

The organisation of the conference was carried out by an Organising Committee constituted as follows: Professor J.G. Hawkes (University of Birmingham), Chairman; Mr. J. P. M. Brenan (Royal Botanic Gardens, Kew); Mr. D.M. Henderson (Royal Botanic Garden, Edinburgh); Dr. F.H. Perring or Miss D.M. Scott (Biological Records Centre, Monks Wood); Mr. R Ross or Mr. T.F.M. Cannon (British Museum, Natural History); and Dr. T. Williams (University of Birmingham).

The Organising Committee was ably assisted by a Secretariat drawn from the Royal Botanic Gardens, Kew: Mr. D. Field, Miss S. Howard, Mrs. H. Hyde, Mrs. S. Kozdon, and Mr. E. Timbs.

After the conference, a brief resume of the proceedings was published in Nature 246: 62 (November 9, 1973) and a fuller account, with abstracts of the papers, in Taxon 23 (1): 101–107 (February, 1974).

As NATO generously offered to publish a full account of the conference, an Editorial Committee (Mr. J.P.M. Brenan, Professor J.G. Hawkes, Mr. R. Ross and Dr. T. Williams) was appointed by the Organising Committee and the present volume is the result.

Thanks are due to the speakers for kindly providing texts of these contributions. Unfortunately three papers have had to be printed in abstract only.

Thanks are also due to the many other people who contributed in a variety of ways towards the smooth running of the conference. I would like to express particular thanks to those who helped in the recording of the discussions: Mr. D.V. Field, Mr. J. Warrington and Dr. W.D. Clayton; also, finally, to my co-editors.

J. P. M. Brenan

CONTENTS

INTRODUCTION

J. G. Hawkes

Birmingham University, England

The purpose of this meeting is to explore areas where Electronic Data Processing can be of help in the storage and retrieval of information concerned with herbarium and other preserved material.

In the scientific world today there is an information explosion which has to quite an extent taken us all unawares. Those of us concerned with taxonomic research have to deal with the ever-increasing volume of literature. Those of us concerned with the organization and direction of herbaria are becoming overburdened with the ever-increasing volume of specimens and the information linked with them.

There is no doubt that in European herbaria, as in those from other parts of the world, we have reached a point when the information content of our collections can no longer be handled by the time-honoured methods of hand-written lists and simple card-index systems.

Nowadays in very many fields computers are helping us to store and handle information in a variety of ways that would have been impossible previously.

In herbaria, too, the time has already come when we need the aid of computers to store, sort and retrieve the data linked to specimens and the ways in which the specimens themselves are stored and sorted. It must be quite clear to all of us here that computers by themselves are useless. Without the proper sorting and presentation of our data and without the proper programming of the computer for handling the data

1

we shall obtain no help from the computer at all.

So, in reality, Electronic Data Processing by means of computers is something we can take for granted as a concept. What is more important is the development of systems of information management. In other words, we must understand and take agreed and concerted action on how to handle the data. This is the area to which our thoughts must be directed.

There are many data management systems available, each with some advantages and some disadvantages. It is clearly most important for us at this conference to discuss these systems and try to agree on the unified use of one of them. In this way, our information will be freely exchange-able between one institute and another, and the methods used can be easily understood from country to country.

It is perhaps worth mentioning here that the management of information by computer-based methods is by no means confined to collections of preserved specimens. Many of us have been concerned for a number of years in establishing unified information systems for what are spoken of as plant genetic resources. By this phrase is meant the living collections of seeds or plants which are preserved for the use of plant breeders and geneticists, and which represent the end products of many thousands of years of evolution under domestication. As an example, work is in progress at Kew and elsewhere to preserve in a living state the seeds of interesting and threatened wild species in seed banks. Computer-based data banks are necessary to store and sort the information on these.

Clearly these seed banks, gene banks or germ plasm collections present the same problems of data management as do the preserved collections which we are to be discussing during this present conference. And indeed, if we can find it possible to agree on the same information management systems for museum collections of preserved plants and animals as we use for collections of living seeds and tissue cultures then we shall have taken an extremely important step indeed.

So much for the general aspects of the problem. I should like to continue with a few remarks on how the organising committee has attempted to structure the programme of this meeting.

In the first place we wanted to try and focus on the problems by asking a number of herbarium directors to say what sort of problems exist and how they would like to see solutions worked out, not in detail, but in general terms. In other words we want to begin by identifying the problems, most of which will be common to all herbaria. It is hoped that these papers will bring out ideas from other delegates related to the sort of information they would like to obtain from their own collections but have been unable to do under present circumstances through lack of electronic data processing methods.

On Thursday we shall be hearing from various speakers about the various information systems they are using and how these are being applied to the information problems they wish to solve.

On Friday the discussion will lead on to other systems with which botanical information systems will need to be compared. In other words we shall look at system interfaces, or compatibility of systems.

Finally, on Saturday morning we shall be engaged in the discussion and ratification of recommendations and proposals which have been sent in to the drafting committee during the conference. In this way we hope that this meeting will provide some very clear and generally agreed conclusions. It is hoped also that it will be possible to reach agreement on a system for general future adoption among the major European institutions. If such agreement can be reached for Europe it is likely to lead on in due course to far-reaching agreements over the world as a whole.

INTRODUCTION

A. Rannestad

North Atlantic Treaty Organisation

Eco-Sciences Programme, Brussels, Belgium

Electronic Data Processing is a tool of considerable importance to modern society. Its full usefulness has, however, often been diminished owing to the incompatibilities of the systems in use. The use of E. D. P. in European botanical collections has, fortunately, not progressed far enough for this to have happened, and it is thus possible to introduce compatible systems. This meeting is therefore very important and timely and I hope that the results will lead to an agreement on descriptors and a basic structure for a suitable E. D. P. system which may be adopted by all major European botanical institutions.

The proposal to call a meeting of representatives from all major European botanical collections, with additional lecturers from American collections and the E. D. P. field, was made by Prof. J. Heslop-Harrison. His proposal was accepted and sponsored by the Special Programme Panel on Eco-Sciences, a subsidiary body of the NATO Science Committee. NATO has extensive programmes in the scientific and environmental fields, and I would like to take this opportunity to give you a short outline of these programmes.

Collaboration and consultation between member countries of the Alliance have been of major concern to the North Atlantic Treaty Organisation ever since it was established. In the mid-fifties a serious attempt was made to implement the collaboration in non-military fields, and a report (1) from a committee of Foreign Ministers, Lester B. Pearson

(1)   Report of the Committee of Three; non-Military Co-operation in NATO. NATO Information Service, B-1110 Brussels.

5

(Canada), Gaetano Martino (Italy) and Halvard Lange (Norway), named scientific and technological co-operation as especially important. As a consequence of this report, a position as Science Adviser to the Secretary-General of NATO (later changed to Assistant Secretary-General for Scientific and Environmental Affairs) and a Science Committee composed of one highly qualified scientist from each of the member countries of the Alliance was established in 1958.

The NATO Science Programmes have changed during the years, but their predominant characteristics have remained with an emphasis on co-operation and catalysis and a capacity for rapid response to new developments. Each of the programmes has been conscientiously designed and deliberately implemented to improve the exchange of information. Over 50,000 individuals, of which some thousands come from countries outside the Alliance, including some hundreds from Eastern Europe, have directly participated in these programmes. The following programmes have been in operation during 1973/74 (1):

> The Senior Scientists Programme   This is a small programme awarding a few Science Lecturerships, Visiting Professorships and/or Senior Fellowships to outstanding scientists.

> The Science Fellowships Programme   This programme, administered by the different member countries, awards about 600 NATO Science Fellowships each year. The programme has allowed about 10,000 scientists to study for about one year in a foreign country.

> The Advanced Study Institutes Programme   An ASI is primarily a high-level teaching activity at which a carefully defined subject is treated in considerable depth in a systematic and coherently structured programme. About 50 institutes are supported each year.

> The Research Grants Programme   The main purpose of this programme is to stimulate scientific research carried out in collaboration between scientists in the member countries of the Alliance. Grants are renewable for up to three years and about 50 new grants are awarded each year.

(1)   More information on the NATO Science Programmes may be found in the booklet Scientific Co-operation in NATO or the book NATO and SCIENCE, An Account of the Activities of the NATO Science Committee 1958-72. NATO, Scientific Affairs Division, B-1110 Brussels.

The Science Committee Conference Programme   The main purpose of these research evaluation conferences is to identify particularly fruitful areas for future research. The recommendations are directed both to those having a responsibility for selecting and supporting research programmes and to the Science Committee itself. One or two conferences are held each year.

The Special Science Programmes   In addition to the general and more permanent programmes listed above, the Science Committee has frequently identified specialised scientific areas as deserving encouragement or preferential support for limited periods. During 1973/74 there are special programmes on: Air-Sea Interaction; Eco-Sciences, Human Factors; Oceanography/Marine Sciences; Radiometeorology; Stress Corrosion Cracking; and Systems Science.

The Sciences Committee Programmes are guided by panels of Scientists from the member countries and support is given to all fields of science, with emphasis on fundamental aspects rather than applications. Results from research projects are published in the literature and scientific proceedings from ASIs and conferences are published in most cases.

I would also like to mention the Committee on the Challenges of Modern Society which, since 1969, has started and co-ordinated pilot studies (1) on: Disaster Assistance; Environment and Regional Planning ; Road Safety; Air Pollution; Inland Water Pollution; Advanced Health Care; Coastal Water Pollution; Advanced Waste Water Treatment; Urban Transportation; and has plans for additional studies on: Disposal of Hazardous Substances; Solar Energy and Geothermal Energy.

(1)   More information on the C.C.M.S. Pilot Studies may be found in the booklet Man's Environment and the Atlantic Alliance NATO Information Service, B-1110 Brussels.

E.D.P. IN MAJOR HERBARIA - THE PRIORITIES

J.P.M. Brenan

Royal Botanic Gardens, Kew, England

Summary

Most herbaria have not introduced E.D.P. methods. It is essential for their advantages and disadvantages to be carefully assessed. An attempt is made to do this with particular reference to the Kew Herbarium. An account is given of the Kew Herbarium and its arrangement. Most herbarium arrangements permit of a more or less efficient information retrieval system to be operated by conventional means and within limits. It is doubtful whether any major herbarium can afford to computerise all its holdings, or whether the result will be worth it. Some suggestions for priorities are given based on needs at Kew. These may be summarised as follows:

1. Inventory of type material

2. Inventories of specimens and geographical areas of outstanding conservation importance

3. Listing of economic uses

4. Listing of vernacular names

5. Recording of vouchers for non-taxonomic research

6. Limited recording of specimens in defined areas, systematic or geographical, of special research interest to Kew

7. Listing of genera and species with their geographical ranges

8. Comprehensive listing of genera and their position under families

I hope that it will be a source of comfort to say that I am aware that a number of delegates to this meeting are not experts in E. D. P. I am certainly not one myself. I feel sure that some at least of the audience must feel rather like the nervous bather, wondering if putting the feet in the water is going to be intolerable or just decidedly unpleasant. I don't want to press the simile, but in both cases uncertainty is the trouble.

We have heard much about E. D. P., but what we have heard has often been rather abstruse, in a language alleged to be English but identifiable often only with difficulty, and not infrequently it seemed to be a sort of evangelism by proponents of particular E. D. P. systems. If curators of major taxonomic collections have felt or continue to feel rather hesitant about the whole concept of E. D. P. let me say that I feel fully sympathetic to their attitude and also that they have not had the informed and expert guidance that they might have expected and which I hope this meeting will go at least some way towards providing.

Initially then the sort of question we must be considering is not "What system should I use?" but "If E. D. P. is introduced what benefit will my institution obtain and how much will it cost in terms of time and money?" I cannot answer myself the last two questions but I am sure that we shall have some concise and informative data later on from Dr. Cutbill of the Sedgwick Museum in Cambridge. Speaking as the curator of a large herbarium, I am sure that we must take a long cool look at E. D. P., assessing what it has to offer and weighing its advantages and disadvantages. Further help may be obtained from Squires (1971).

In planning the opening contributions to this meeting the Keepers of the three major national herbaria in the United Kingdom, Mr Ross of the British Museum (Natural History),   Mr Henderson of the Royal Botanic Garden, Edinburgh, and myself attempted to carve up the problem between us, in order to prevent tedious repetition, and it has fallen to me to speak on the priorities. In other words, given that E. D. P. methods have at least a useful contribution to make in herbaria, as I am sure that they have, what needs to be done first, and what are the most important jobs to be tackled?

I shall naturally use the herbarium at the Royal Botanic Gardens, Kew, as my point of reference, but before going further let me give some relevant information about the Kew herbarium.

Firstly, its size. Some wildly varying guesses have been made in the past. As recently as 1959 an authoritative estimate appeared in print of "some 6 million ... specimens" and this was subsequently increased, though I believe only verbally, to 7 million. In 1967 a move of a major part of the herbarium to a new wing offered a good opportunity for a more rigid test based on controlled statistical sampling. The results, surprisingly, indicated a total number of specimens in the whole herbarium of between four and five million. This, I must emphasise, referred only to specimens mounted on conventional mounting sheets, with allowance made for multiple tenancy of single sheets (Brenan & Carter, 1972).

The herbarium is world-wide in coverage, though not systematically comprehensive. As a result of a division of labour between Kew and the British Museum (Natural History), officially agreed in 1961, responsibility for research on certain groups, notably the bryophyta, algae and lichens, was assigned to the Museum, and the Kew holdings of these groups have been transferred to the Museum on a basis of permanent loan.

A number of ancillary collections are also present. For example, there is a collection of more than 30,000 specimens in jars of liquid preservative, mostly fleshy or delicate flowers and fruits. There is also an extensive carpological collection, a seed collection, a pollen-slide collection, and, last but by no means least, a collection of more than 150,000 drawings, engravings, paintings and photographs of plants derived from a variety of sources. In all these collections there are cross-links, sometimes multiple. For example, a specimen mounted in a herbarium sheet may also be represented in the carpological collection, and the spirit collection, etc.

It must also not be forgotten that Kew is not only a herbarium, but a botanic garden and also carries out research in a variety of botanical disciplines. Specimens in the herbarium may then also be vouchers for cultivated plants in the gardens or for research carried out in the Jodrell Laboratory. We can thus see that the herbarium carries quite a complicated cross-reference system, purely for use within Kew. I will leave it to others to elaborate the wider inter-institutional or international cross-reference system that is inevitably built into every major taxonomic collection.

This, briefly, is the body of material and data we are dealing with at Kew, but doubtless many other major herbaria can paint a similar picture. For any reasonably efficient method of retrieving information and data from 4-5 million items a carefully planned system of arrangement is essential. I wish to emphasize the point here that the problems of data and information retrieval, though perhaps they have not been long known by so clinical a title, have long ago been realised by herbarium curators and their difficulties at least partly faced.

In other words most herbaria are already arranged to allow of some more or less efficient method of information retrieval. E. D. P. is a new method of solving an old problem and its effectiveness must be judged on a comparative basis.

Again referring to Kew, what is the system of herbarium arrangement? It is primarily systematic, down to the level of family and genus. All material of a given family or genus is together, other than what is in the ancillary collections. There is an index to families and their location, the latter given by a combined number and location code for each block of cupboards. For example a family location might be indexed as B 23, meaning that the family starts in cupboard-block number 23 in B wing of the herbarium buildings.

Within each family there is a systematic index to genera, each genus being numbered. The location of the generic index is colour-coded by a red label on the outside of the cupboard containing the index.

The arrangement of families and genera is by a modified version of Bentham & Hooker's Genera Plantarum. The objection is sometimes made that the system is neither modern nor phylogenetically up-to-date. This is to miss the point. The arrangement is primarily a system allowing a reasonably quick retrieval of information based on identified and localised specimens and the relative order of families is in comparison of much less importance. A herbarium does not (and probably cannot) attempt to display any phylogenetic system.

At the level of the genus, geography in general takes precedence over taxonomy. The world is divided for this purpose into a series of numbered regions and subregions. For example Europe is 1, India 5, Australia 7, Tropical Africa 10 but divided into 5 subregions, 10A, 10B etc. Each region and subregion is thus coded by a combined system of numbers and letters, and each continent has its regions and subregions indicated by labels on the genus covers in the herbarium of a distinctive colour for each continent.

Within each region or subregion the arrangement of the species is systematic, normally following acceptable recent revisions on floras, and only as a last resort alphabetical by specific epithets. It may thus happen that the arrangement of a genus may differ from region to region, or even between subregions, if different but acceptable recent revisions have appeared for different subregions. To make inter-regional comparison easier more general systems are preferable but too rarely available. Indices to species are provided where necessary.

Often the sequence of countries within a given region or subregion follows a standard sequence and this may even be carried further.

In other words the specimens are filed in such a way as to take account of both systematics and geography, and similar systems, of greater or less sophistication, probably apply in most major collections.

A very useful convention has arisen over the years whereby collectors number their gatherings serially, without repetition. <u>Brenan</u> 11043, for example, denotes an individual gathering on a given day, at a given locality, of one taxon. Confusion can of course arise when there is more than one collector with the same name. In general, however, given a collector's name and number, the country of collection, and the name of the taxon, it should be possible to retrieve a wanted specimen with its associated information in a matter of minutes, provided it is there!

I have perhaps dwelt in too much detail on the minutiae of herbarium arrangement, but I wish to make the point that the herbarium may be compared with a card index and that it has its own inbuilt methods of information retrieval by conventional means and that these, within their admitted limitations, have proved not inefficient.

It is important to define what can or cannot be done by existing methods, for otherwise it is difficult to make comparisons with alternative methods of retrieval by E.D.P. and to see the advantages of the latter.

I remain somewhat sceptical of the claims made by Shetler (1969 pp. 730-736) that E.D.P. in the Herbarium can control loans and exchanges and "herbarium transactions" in a way so much more efficient that existing methods can be superseded. Nor am I convinced of the utility of recording by E.D.P. all new accessions, but neglecting the backlog.

The main merit of E.D.P. methods lies in the possibilities of rapid selective retrieval of data or information in ways that are independent of or cut across systematics or conventional geographical divisions. It may be argued that the existing systems satisfy the needs, but the limitations of the existing systems are well known. If questions are put of the following sorts, for instance: what species are recorded from localities above 10,000 feet altitude in the Peruvian Andes? or, what plants in the Amazonian region are recorded in the herbarium as being of medicinal use?; then an answer is impossible by conventional means in the herbarium without excessive expenditure of time and labour, and

such questions are usually just not put. For general comments on this
matter see Meadows (1973).

Greene (1972) estimated that a competent machine operator can
handle 750-1000 cards per month, working 5 half-days per week. Given
that a complete recording of 4-5 million specimens is just not on because
it would prove far too demanding in terms of manpower and time for it to
be completed in a reasonable period, what might be the priorities for
E. D. P. ?   At this point, I must emphasize that an E. D. P. system in the
herbarium can serve a useful domestic purpose, but its usefulness is
enormously increased if it becomes part of an automated inter-institutional,
international information system. Now for the priorities as I see them:

1) Types   As all working taxonomists know, type specimens in their
wide sense are not just items of historic or antiquarian interest, but vital
documents for current research. By the provisions of the International
Code of Botanical Nomenclature the application of the scientific name of
a taxon is controlled not by a description but by the identity of the type.
Duplicates of types (isotypes) may be only of slightly less importance
than the orginals (holotypes). Types, in their wide sense, are scattered
through the herbaria of the world, but no central register of types or
their location exists. An effort has been made by the Smithsonian
Institution in Washington to form a central index of American types
(Shetler, 1969), but this initiative has not been followed by other continents.
For Kew I would say that a register of the types (some 300,000; Brenan
& Carter, 1972) is a most important priority for E. D. P. treatment.

2) Conservation   The current timely interest in the conservation of
genetic resources has shown the value of major herbaria as sources of
information. For some species now apparently extinct the herbarium
material provides the only verifiable link with the appearance and features
of the plant as it once lived. Furthermore an analysis of selected
specimens in the herbarium may provide the fullest record and the
easiest means of recording species genuinely endemic to areas of
conservation importance. Conventional methods of herbarium information
retrieval are not very helpful, but E. D. P. methods, particularly as part
of an international scheme for the storage of environmental information,
certainly would be.

3) Economic Uses   Label data in the herbarium are a rich source
of references to economic uses of plants. There can be no doubt that
many indigenous uses of plants are of potentially wide value to humanity
but have not been tested. Even inventories of known economic uses of

plants within countries or areas are few and often imperfect. The information is scattered and very difficult to retrieve except by laborious examination of labels, specimen by specimen. The evidence is there but mostly untapped. The revision of Dalziel's Useful Plants of West Tropical Africa, at present being undertaken by Mr H. M. Burkill was deliberately based on the Herbarium and Library at Kew.

4) <u>Vernacular names</u> In tropical countries particularly, languages may be numerous, complex and local. To the man in the field, agriculturalist or forester, the use of vernacular names and a clear understanding of the scientific identities of the plants to which they refer may be vital. The herbarium is particularly important as a source of information here, as the records of vernacular names on specimen labels are very numerous and they are correlated with actual plant specimens whose identity is usually subsequently verifiable.

5) <u>Vouchers for non-taxonomic research</u> The validity of all research involving plant materials depends on the identity of the research material used being initially ascertained and later verifiable. Alas, the condition is often not fulfilled. How often in the past has the importance of making voucher specimens been forgotten. A computerised register of important voucher specimens and their location would seem to me highly desirable. For the taxonomic purposes of the herbarium, it would be useful to have access to a register of vouchers, for example, of cytological and biochemical research.

6) <u>Areas of special research interest</u> Each herbarium has its areas of specialist research interest, whether systematic or geographical. These provide and have provided opportunities for recording for mechanised retrieval defined but reasonably complete sectors of herbarium information, which may thus become quickly and generally available. I am convinced that reasonably complete recordings of limited fields are vastly preferable to incomplete coverage of wider bodies of data. A very good example of E. D. P. methods applied to a small herbarium of some 30,000 specimens for a clearly defined objective, in this instance the British Antarctic Survey herbarium, was given by Greene (1972).

7) <u>Geographical Survey of Genera and Species</u> I am indebted to my colleague Dr. R. K. Brummitt for suggesting a scheme for recording genera and species represented in the herbarium at Kew, together with the countries from which each is represented by specimens. If this information is available in machine-readable form and retrievable selectively, then a very significant body of information would become

available for such purposes as phytogeographical and chorological studies, for the production of check-lists of plants occurring in countries, for obtaining distribution information to serve the needs of conservation studies, etc. The countries recognised would be standardised by a prearranged system. Although the result would be imperfect in that genera and species unrepresented at Kew would be omitted, yet the results would be a sufficient approximation to the truth for their value not to be in question.

8) Standardised Family and Generic Catalogue   The herbarium at Kew in its arrangement of genera under families represents the cumulative result of many expert judgements and studies. This has already given rise to an agreed list of recognised families for internal use at Kew and the British Museum (Natural History). Were this to be extended to genera with the agreed family position it would serve as an unified system by which all the various taxonomic collections at Kew could be arranged. I must emphasize that the importance of this would be mainly domestic. Other institutions have their own systems and experience has shown that a search for general agreement over general classification for families and genera is likely to be a lost cause!   Nevertheless the principle may be generally useful. I am grateful to Mr D. R. Hunt for his suggestion over this matter.

It is essential to look carefully at the cost-effectiveness of E. D. P. in the herbarium. It is not cheap, so the advantages must be commensurate. I have tried to indicate some of the priorities bearing this in mind. I shall be happy to receive suggestions for additions or comments on the items included from directors and curators of other herbaria present here.

REFERENCES

Brenan, J. P. M. & Carter R. G. (1972). The counting of the Kew Herbarium. Kew Bull. 26: 423-426.

Greene, Dorothy M. (1972) A taxonomic data bank and retrieval system for a small herbarium. Taxon 21: 621-629

Meadows, Harriet Krauss (1973). The use of generalised information processing systems in the biological sciences. Taxon 22: 3-18

Shetler, Stanwyn G. (1969). The herbarium, past, present, and future. Proc. Biol. Soc. Washington 82: 687-758

Squires, Donald F. (1971). Implications of data processing for museums. Cutbill, J. L., Data Processing in Biology and Geology: 235-253

THE DATA FROM HERBARIA

D. M. Henderson

Royal Botanic Garden

Edinburgh, Scotland

Summary

The lecture assumes that the intention is to consider an E. D. P. system capable of holding all the useful data in European herbaria but excluding the specimen data necessary for numerical taxonomy. It will examine the types of data available on sheets, their validity and the problems they raise. The general suggestions are:

1. A uniform family system could be adopted easily for the phanerogams but some cryptogamic groups present problems.

2. The lack of universally applicable generic systems and the poor state of revision of herbaria present major problems for input at generic, specific and subspecific levels.

3. The volume of correction following input of collections as they are at present would be very great. Continuous revision of herbarium naming would require continuous input and would inhibit normal curatorial work. Thus the desirability of computerising data only after major revision must be considered.

4. Geographical data will present major difficulties especially with old, poorly localised collections. If any automated geographical print-out is to be possible, the compilation of special gazetteers and the conversion of geographical information to a standard grid system might well be obligatory.

5. Authorities for plant names could not be reliably copied from existing
    labels. The input of plant names if authorities are added would
    involve a major nomenclatural exercise covering the whole plant
    kingdom. It might be advisable to dispense with this input as there
    may be no good use for its output. It would parallel a revised
    Index Kewensis which would be more informative and valuable.

6. Ecological data on herbarium sheets is not necessarily sound.
    Codifying it would increase inherent error but the inclusion of this
    information might be marginally worthwhile.

7. A standard list of plant collectors would be required. The problem
    is not great with major collectors, but minor, often local, collectors
    would at least double the task for very little return.

The general conclusions are that input of the data as it stands in
herbaria is possible but major revision before input would be highly
desirable. This prior revision must be added to any estimates of the
work load that the project involves.

———————

This contribution attempts to survey the data available in herbaria
and to assess how much of it may be worthwhile attempting to process.
It was designed to provide a framework for discussion and argument,
rather than an exhaustive survey of the details of the subject. Many
points still remain imprecise because they can only be assessed by an
adequate pilot scheme - estimates and guesses can be very inaccurate
as Shetler has recently stressed in his account of a type specimen
register for North America. It is perhaps worth establishing at the
outset that consideration is restricted entirely to information available
on herbarium sheets, with the tacit assumption that only data retrieval
is being considered, and not systems of numerical taxonomy.

The curator of any large herbarium is very likely to see the problems
in an electronic data processing approach rather than the possible
advantages. This is perhaps because the problems do exist, but the
advantages are somewhat theoretical and not always of direct application
to herbarium taxonomy itself. Thus, only a pragmatic approach to the
problem may be satisfying to curators.

In preparing this contribution, the subject was approached by taking
a few covers from the Edinburgh herbarium and looking critically at the
information on them. It is simplest to consider the information under a
number of headings almost as they occur on the herbarium labels.

To take the problem of a list of families first, one system can be adopted - and for retrieval probably must be adopted. No one taxonomist will be necessarily happy with every decision, but surely it is not beyond the bounds of possibility to get some generally acceptable system. Certainly, it would not be too difficult in the Angiosperms where families are well established; but this is not necessarily so in the Cryptogams. A system is fairly well elaborated for the Bryophytes, but this is certainly not so for the Fungi. Even where there is a fairly well established system, many genera are of uncertain position within it. If we consider generic and specific names, as they are in herbaria at present, more serious problems arise in large world-wide herbaria. Perhaps herbaria have themselves to blame if outsiders believe that the contents of herbarium folders arranged in families, genera and species are in good order, perfectly named according to the best taxonomy available. But it is not so. Nearly every large herbarium has considerable sections, either taxonomic or geographic, which have hardly been revised in the last fifty years. The reasons for this are well known, and include uneven taxonomic treatment because of regional monographs or floras, and lack of staff to curate collections and keep them up-to-date. But if we delay until the herbaria are in perfect order, no living taxonomist will be extant, for certain complicated problems, such as a worldwide, uniform treatment of the Compositae or the Rubiaceae, cannot be achieved within the foreseeable future. Thus, the input of unrevised names must be accepted or the input from these difficult groups would have to be postponed indefinitely.

The problems posed by subspecific ranks are very similar to those of higher taxa, except where a multiplicity of lower ranks - subspecies, variety, form, subform - have been used. Some simplification for input would be desirable for otherwise retrieval would pose major problems. It would be necessary also to make provision for cultivar names. Some subspecific names pose problems, but their inclusion in an E.D.P. system would give a facility for working compilations of subspecific names - a facility which is not available at present and whose lack produces problems for the working taxonomist.

Consideration of plant names leads automatically to author citations. The proposed Kew list of authors' names would certainly help to standardise input. Nevertheless, it is doubtful if authorities could be copied straight from herbarium sheets. Old sheets in less worked parts of herbaria are sadly out of line. Input would require bibliographic work, which is perhaps not worthwhile. The question must surely be asked whether the input of authors' names would achieve any useful purpose.

The work requires to be done <u>once</u> for a new Index Kewensis. It should
not be done repetitively for the input to an E. D. P. system.

These questions of the names on herbarium sheets have been dealt with
at some length because curators are most doubtful about the validity of
these data. The output of an E. D. P. system would undoubtedly expose
many of the faults in input and allow its correction. The critical question
in arriving at some conclusion here is: would greater resources be
required before input or after output to correct the data? This is surely
a question to be answered by a well-conceived pilot scheme. Of course,
many sheets bear multiple determinations. Dr. Greene of The British
Antarctic Survey included all the names ever appended to specimens in
his system for Antarctic herbaria. Retrieval of this information gives
lists of synonyms and mis-identifications. An E. D. P. system for
major herbaria must surely omit this refinement. An exception must
be made for types where input of the basionym would probably be necessary
in addition to the current binomial.

After names comes geography. Some of this is simple. Altitudes can
be converted to one unit, but site localisation presents much greater
problems.   Unless the locality is fed in with some artificial numerical
finder system - latitude/longitude or kilometre grid square - output is
not possible on a geographical basis unless one is prepared to accept
all the difficulties of impermanent political boundaries. Conversion to
any artificial system would be tedious. Certainly, only, at most, 10 per
cent of the specimens in the Edinburgh herbarium have latitude/longitude
co-ordinates. The arbitrary grid system, as used in Europe, would
impose a similar task before input. It must be recognised also that it
is quite possible to add a mechanical system to a specimen which is
fairly accurately localised, but many of the early collections and, perhaps
most important of all, many type collections, are those which are least
accurately localised. Without over-emphasising some of the difficulties
which would undoubtedly arise, the specimens, sometimes type, from
cultivation, must not be forgotten. Even recent collections can pose problems
which are very time-consuming. It is apparently necessary to use
Kingdon Ward's travel books to locate many of his collecting sites. It
may be that the preparation for input of geographical data would require
the prior compilation of detailed gazetteers; but even gazetteers will not
solve all problems. The Flora of Turkey group in Edinburgh have worked
intensively on that country for a dozen years and have a working gazetteer,
but still they experience problems involving long searches. This problem
of geographical localities must be assessed by a realistic pilot-scheme if
a time-wasting counter-productive effort is to be avoided. Major problems
with even 0.1 per cent of a million specimens may be the sort of measure

of the problem. There is one converse argument. The chronological print-out of any collector's collections will immediately help to narrow the area of search for any additional collections which are at least dated. Thus, once an E. D. P. system is attained, it could go a long way to remove the present impediment to finding localities.

So far, the information is that used regularly by taxonomists. The ecological data which appears on some sheets forms a somewhat different category. In the first instance, it is perhaps wise to remember that most plant collectors are not ecologists. Furthermore, we must bear in mind that although the taxonomic system is somewhat chaotic, ecological classifications are even more so, and there is certainly no world-wide detailed system which could be applied. Perhaps the broadest of vegetational categories - savannah, evergreen forest, tundra, etc.-might be used and the information squeezed into these broad groups. Certainly, if the data is to be retrieved selectively according to ecological categories, then the information must surely be so ordered. Information on herbarium sheets as regards ecological association is so haphazard that input would surely require interpretation of the information with all the risks of misinterpretation. Before we too readily embark on codifying ecological information, perhaps we should remember that our herbaria have not been amassed for ecological purposes.

As an additional complication, we should remember that ecological information can involve problems of language - a problem which is absent in the other categories of data that we have examined. Although this consideration of ecological information is obviously chiefly concerned with Angiosperms, there are many other problems in other plant groups. These include the hosts of parasitic taxa, especially in the cryptogams, the zone information for marine algae, the saxicolous or epiphytic habitats of bryophytes and lichens. The data here is usually accurately noted and certainly worthy of record. It might be thought that the input of collector and date of collection should pose no problems. A decision would be necessary in relation to the form of date to be used, but this could be done without a new world/old world confrontation. The collectors' names would probably require a standard catalogue, as for authorities. Another problem arises, however, in considering the great exsiccata with their multiple, often minor, contributors and the innumerable local, very minor collectors. The recognition of all the major "Smiths" who have collected plants is sufficient of a task without adding all the minor "Smiths".

Collectors' numbers may also cause some difficulty. Bornmüller used a different series in different years. Ultimate confusion would be

avoided in an E.D.P. system in that computer recognition of individual
sheets almost certainly implies individual numbering of every herbarium
sheet. It must be noted that perhaps half the major herbaria do not
number their sheets serially at the moment, and that serial numbering
would produce inconveniently large numbers with up to eight or nine digits.

Turning to other items of information, the problem of types arises.
Taxonomists are fairly well agreed that some list of types and their
location is highly desirable. An E.D.P. effort on herbaria could meet
this need. If herbaria were perfect, all types would be annotated and the
category, holotype, isotype, etc. stated. But between 25-50 per cent of
all types in British herbaria probably have no indication whatsoever of
their type status. Moreover, if input were carried out without
surveyance of every sheet by a highly knowledgeable taxonomist, possible
type status will not be detected. Furthermore, even if the possibility
that a specimen is a type arises, a time-consuming bibliography check
would be required to confirm this. Therein lies a dilemma. Input at the
purely clerical level would miss the types; the effort and time of skilled
staff to detect them may be too great.

Nevertheless, it can be maintained that some of these problems would
be aided once a system began to operate. Types entered for one institute
would automatically help to identify types in others. This leads one to
suggest that an intensive effort on particular collections in certain
institutes would be the most effective method of proceeding. To turn
from the general to the specific, it would be foolish if all institutes
duplicated work on George Forrest collections when the main set in
Edinburgh could be the primary source of information.

There remains a further great mass of miscellaneous information:
common names, economic uses, flowering time, growth habit, state of
the collection (whether vegetative, flowering or fruiting). We must
balance the desirability of having this type of information against the
usage of skilled staff to assess the herbarium sheets for input and the
extra input time. Herbaria must surely ask the question whether the
system is being elaborated for herbarium purposes or are they producing
a system for others. When this problem has been considered it must be
borne in mind that for a primary input all sheets would probably be
handled once. So it would be clearly unwise not to extract all the
information that we can foresee being used in this one operation.

This leaves a number of miscellaneous data problems which must
surely be treated with a sense of proportion. All curators know of the
problems of the indecipherable writing, the completely inadequate label,

the undetermined specimen. The first two are difficult, but relatively
minor in numerical importance. The third - the undetermined specimen -
raises a question of up-dating the data. At least we may be glad the
herbarium sheet is relatively immortal compared to the living plant in a
botanic garden. We do not have the problem of continually moving
localities, but we do have the problem of re-determination of sheets.
All re-determinations would have to be recorded for input and this would
be a serious and continuing burden for herbaria. All loans would have
to be checked on return. Perhaps even more serious would be the
repercussions that formalising re-determinations might have on the
casual up-dating of our herbaria which proceeds all the time. Much
minor re-determination goes on and this might cease were it to be
done with greater deliberation, and hence at the expense of many more
man-hours. Informal curating might cease, to the detriment of the
general state of the collections.

From this very general survey there seem to remain two points
which appear to be worth re-emphasising. We have recognised the
great inaccuracy of some of the data and the problems that correction
of it would pose. But the fact must not be lost sight of that if the best
curated collections, defined by taxonomic compass, by geographical
area or by collector, were fed in first the preliminary output would be
a valuable tool in correcting further input. Indeed, one could claim that
there might result a very considerable upgrading of the quality of
curation of our collections as a result of an E. D. P. exercise. This
does not, however, lessen the magnitude of the task considerably.

The second general consideration arises from the fact that I have
somewhat simplified the subject. This has been done deliberately, and
in the light of experience in the Royal Botanic Garden, Edinburgh, of
planning and completely implementing an E. D. P. system for the living
plant collections. This task could not have been achieved had we not
limited the number of our objectives and faced and solved the many
troublesome, but sometimes numerically minor, problems without
disturbing the general overall system. From this point of view, it would
be possible to do major collections with a limited number of descriptors
for input. This opinion is based on a comparison in that the Edinburgh
herbarium collections are approximately 100 times the size of the living
collection. Nevertheless, I am convinced that if we produce too
complicated a system, with too many descriptors, we will never achieve
anything worthwhile in our lifetimes.

DISCUSSION, MORNING, WEDNESDAY 3 OCTOBER

This discussion was a general one on the papers read during the morning.

Mr. R. J. Pankhurst - The data which we may want to put in is variable both in its nature and the ease with which it can be obtained. Instead of a total approach in which all the information is put in, we should plan for partial input because it is easy to obtain additional information for later incorporation as a programme of consolidation. The standardisation of the meaning of contents of data bases is much more important than standardisation of its format. This is because format can be transformed automatically if necessary, but it is much harder to alter the meaning.

Mr. D. M. Henderson - I agree. There are many ways in which this could be done. Data from new revisions of groups could be added as could geographical information from areas on which work has recently been done.

Mr. G. Harrison - Would it not be useful to show negative information - what is not available?

Mr. D. M. Henderson - This would always show up as a blank in the standard print-out.

Mr. J. P. M. Brenan - I agree. It is essential to use the simplest method of recording information.

Dr. D. B. Williams - There should be a standardisation of place names. Many specimens have localities of archaic origin. Where there has been a positive addition of non-original data interpolated, this should be shown clearly.

Dr. J. L. Cutbill - In considering the economics of about 1000 records, the bulk of the cost is in the physical handling of the specimens or data; there is comparatively little in getting this information into the computer.

Professor C. H. Oppenheimer - The selection of systems must be up to the user of the information. An open-ended system is capable of having additions made to it or subtractions or ignoring as well as identifying negative information. It is necessary to list a group of descriptors in the form of a dictionary which would identify how we would process the specimens. A common base line of descriptors would be the common names, scientific names and geographic localities.

Mr. J. P. M. Brenan - The sort of information is more important than the way it is put in. There could be some difficulty over the assessment of types but the subsequent refinement of these is a matter which should be left to the monographer and should not concern the cataloguer. Old ecological data present a problem but we could come to an agreement on descriptors for use in future ecological work.

Professor J. G. Hawkes - We should not hold back from using E. D. P. for groups which have not been worked on recently. The information from E. D. P. is of the greatest use in revisions. Any system is capable of being edited or updated as more information becomes available. An alteration of the classification is quite possible and we can add a special coding for the degree of reliability of our data of doubtful significance and we can also include the date when the material was used in research or seen by the last monographer. We should utilise what information we have but assess its relative value.

Professor K. Walther - About 75% of our enquiries concern types and data processing of these would seem to be of the greatest importance.

RELATIONS BETWEEN HERBARIUM RECORDS AND OTHER RECORDS

R. Ross

British Museum (Natural History)

London, England

Summary

In considering the design of E. D. P. records of herbarium holdings it is necessary to take account of the need for compatibility between such records and those of other types of collections, viz. :

Associated material, such as spirit specimens, pollen specimens, carpological and other bulky specimens and anatomical preparations ;

Other botanical collections, including collections of living plants, collections of microscopic organisms, and palaeobotanical specimens;

In some institutions, zoological collections.

The various points that need to be borne in mind in connection with each of these are discussed.

The papers by Brenan and Henderson that precede this one have set out some of the considerations that need to be borne in mind when planning an electronic data-processing system based on herbarium specimens. They have, however, confined themselves almost entirely to the handling of the data on the herbarium sheets themselves and their labels. In so doing they have made the point that the electronic data-processing systems used in different herbaria should be so designed that it will be possible to pool the records of the different herbaria with the minimum

of difficulty, thus enabling the production of a data base that will be
comparable to Union Catalogues of library holdings, indicating what
material exists in herbaria as a whole and in which herbaria any
particular gathering is to be found. Compatibility is required, however,
not only between the records based on herbarium sheets in different
institutions but also between those records and others based on different
types of material. This, the third of the introductory papers to this
conference, lists the most obvious of these other types of material that
will, or may, need to be taken into account when an E.D.P. system for
dealing with herbarium specimens is being planned and draws attention to
some of the features of the records of such material.

The need for this compatibility arises for two reasons. There are
some types of material that need, when present, to be examined together
with the relevant herbarium sheets when taxonomic studies are being
carried out, and other types that, whilst used primarily in independent
studies, may nevertheless also be relevant to research based primarily
on herbarium specimens. In such cases cross-references between the
herbarium specimens and these other types of material, particularly in
so far as they are derived from the same gathering, will be necessary.
In addition there will often be, in the same institution as an herbarium,
collections not necessarily related to herbarium specimens at all.
Economy requires that data about these should be capable of being
handled by the same equipment, and, as far as possible, using the same
programs.

The enumeration that follows begins with those types of material
most closely associated with herbarium specimens and hence with:

Spirit Collections   These supplement herbarium specimens in those
cases where drying destroys information about the morphology of specimens,
e.g. the flowers of Balsaminaceae, Commelinaceae and Orchidaceae.
They are almost invariably part of a gathering that is also represented
as a herbarium sheet. As these specimens are not normally studied
except in conjunction with herbarium sheets, a separate record of them
may not be necessary; all that may be required is an annotation in the
herbarium record that spirit material exists, and possibly an indication
of where it is stored. Tubes and jars of spirit material are less easily
re-arranged than herbarium sheets and may be stored in a static and
arbitrary sequence, rendering this latter information necessary.

Box Collections   I use this term, rather than carpological material,
since not only fruits and seeds, but also such items as fern rhizomes
and bracket fungi, may be too bulky to be mounted on herbarium sheets.

Some of the items stored in boxes can be regarded in the same way as spirit material, for they too are normally only studied together with herbarium specimens; fern rhizomes are an example of this category. Others, including coralline algae and bracket fungi, are substitutes for herbarium specimens, and data about these will need to be included in the herbarium record, with an annotation of their nature and location. Carpological material, on the other hand, whether boxed or in the form of microscope slides of small seeds, is used partly in conjunction with the corresponding herbarium specimens and partly as the basis for independent studies, such as the preparation of manuals for seed identification. Because of its autonomous use it will probably be considered desirable to have a separate record of such specimens, cross-referenced to the herbarium record. The data recorded about this carpological material will be a repetition of that in the record of the corresponding herbarium specimen plus any necessary information about the nature of the carpological specimen: fruit or seed, boxed material or microscope slide, and location of specimen.

Whether or not the data common to the records of the herbarium specimens and the carpological specimens will need to be stored twice in the computer will depend on the hardware and programs available.

Pollen Collections   Pollen is studied autonomously to a much greater extent than seeds but it is nevertheless examined in herbarium-based taxonomic studies. It will accordingly need a record similar in type to that required for the carpological material. However, pollen from a single gathering may be stored in a tube and as a microscope slide, and the sample, or part of it, may be acetolysed, and the record will need to contain data on these aspects.

Anatomical Material   As far as vascular plants are concerned, this type of material is almost entirely the subject of autonomous studies, even although these are often taken account of in taxonomy. Nevertheless voucher material is normally preserved in the herbarium and cross-referencing to the herbarium record will be necessary here too. The data to be recorded about anatomical preparations, over and above that common to the herbarium record, will need to include the part of the plant represented by the specimen, the direction of the section, and the technique used in preparing the specimen.

Microscope slide preparations of some groups of lower plants such as the bryophytes and the larger algae, however, are as closely related to the herbarium specimens as the spirit material of higher plants and, as far as records are concerned, can perhaps be dealt with in the same way.

Living Collections   This heading, literally interpreted, would include culture collections of microscopic organisms, but these are treated separately below.   In a later contribution to this symposium Cullen (pp. 167-175) describes a record system for a living collection. This indicates the kind of data that needs to be recorded for such a collection. Where a living collection and an herbarium both form part of a single institution, an attempt is normally made to preserve in the herbarium voucher specimens of the plants in cultivation and hence cross-reference between the two records will be required.   Their content will not be identical but it is highly desirable that the provenance data in the records of the living collection, e.g. whether the plant was transplanted from the wild, grown from seed collected from wild plants, or received as stock or seed from another garden, should be associated with the herbarium specimen.

Culture Collections   For the most part these consist of organisms for which no satisfactory method of permanent preservation is available. They are nevertheless maintained in some herbaria where taxonomic work on the groups concerned is carried out and when this is the case the recording system adopted will need to deal with them, and hence it must be designed to deal with such items as strain numbers, provenance and culture medium used.

Microscopic Plants   Collections of at least some groups of these are organised in a very different way from herbaria.   To take the diatoms as an example, the main collections always consist of microscopic slides, and these are of three types:

Single-species preparations - These are made by the picking out and mounting of individual specimens.

Arranged preparations of many species   - In these, specimens of different species, usually from the same locality, are individually picked out and arranged in rows, so that each specimen has a definite position that can be specified by row and number.

Strewn slides - In these an unsorted sample of the population in a a particular gathering is mounted on the slide.

Since both the second and the third type of slide usually contain specimens of a number of genera from different families, the collection cannot be arranged in systematic order like an herbarium.   Also, it is quite impossible to record in the limited space of a microscope label the data about the provenance of the material and all the identifications, which may be more than a hundred for a single slide.   As a consequence the slides that make up a collection are normally numbered in a continuous sequence as received, and documented by a register in which details of each slide, including locality and collecting data, type of slide, mounting medium, and preparer, as well as the identifications, are recorded. A species index is maintained in addition. The generation of this from a machine-readable version of the register is an obvious requirement for a data-processing system.

As well as the microscope slides, the collections include both the raw samples, in tubes or, where the material is fossil, boxes and also tubes of material processed ready for mounting, and the records of these will need to be cross-referenced to the data about slides from the same sample.

Palaeobotanical Material   Collections of this type will require in their records information under various heads that are not applicable to recent plants, as well as others that are. The principal ones special to fossil plants will be: details of the rock formation in which the fossil was found, both stratigraphic and lithological ; the mode of fossilization, whether an impression, a cast, a petrifaction; and the part of the plant preserved, whether root, stem, leaf or fructification. The great variation in bulk of plant fossil specimens, and the fact that in their preparation they may be divided and sectioned, results in a need for a more elaborate system of recording their storage location than is needed for most other sorts of material. · Furthermore, geologists do not normally adopt the system of labelling gatherings by collector's name and number that is used by botanists, and the data-processing system will need to cope with a different method of tagging specimens from that used for recent plants.

Zoological Collections   Some herbaria are parts of institutions that cover the whole field of natural history, and these will wish to use a system for data-processing that can be applied to the whole of their collections. Zoological collections are even more varied than the whole range that we have considered so far. Protozoa mounted on microscope slides pose similar problems to diatoms; nematodes and other lower metazoa may be prepared as long series of serial sections, or may be stored in tubes in spirit, the tubes themselves being stored in jars; insects on their pins can be arranged and kept in systematic order much like herbarium specimens, as can the shells of molluscs ; fish, amphibia, and reptilia are normally stored in jars of spirit but may also be preserved as skeletons; specimens of birds and mammals are skins or skulls or skeletons, and occasionally spirit specimens. Details of sex are almost always needed and in many groups the stage in the life cycle must also be

recorded. The types of habitat description appropriate to different
groups of animals also vary greatly, and are often not those normally
applicable to plants.

Another point that must be borne in mind is that the geographical
system adopted must cover the whole surface of the earth and not be
confined to the land and immediately adjacent waters, which is all that
is required for most plants. For many marine organisms localities
will be specified as the station number of the cruises of research ships
and this is another item with which a comprehensive record will have to
deal.

Other Material   All the items considered above have been specimens.
Many herbarium collections are, however, supplemented by collections
of illustrations, both drawings and photographs. An adequate record
will need to include information about these, especially those that are
based on specimens in the herbarium.

This rather cursory and certainly incomplete listing of the types of
collections additional to herbaria, and the data associated with such
collections, makes clear certain characteristics that any system for
automatic data retrieval adopted in an herbarium must have.

(i)   There must be no constraints on the choice of the types of
       information (descriptors in the sense of Rogers, see page 122)
       to be included in any particular data bank. Thus it must be possible
       to decide for each type of collection, e.g. herbarium specimens
       or slides of diatoms, the set of descriptors under which the data
       for each item is to be entered. Within the data bank for any one
       type of collection the descriptor set will need to be uniform through-
       out but it must be possible for it to vary between one type of
       collection and another.

(ii)   The entries must be free-field. It is impossible to predict in
       advance all the entries that it will be necessary to make under
       many of the descriptors, and if this cannot be done it is impossible
       to develop the system of coding required for fixed-field entries.

(iii) For the data bank based on any particular type of collection it must
       be possible to choose the particular descriptors by which information
       can be retrieved from the data bank, and to specify what other
       data about each item is to be provided in the output. For example,
       from the data bank based on herbarium specimens it must be
       possible to obtain a listing of all the data about all the specimens

of a particular species, or to get a list in numerical order of all the specimens of a particular collector with only their identifications. In this connection it must be borne in mind that it is only possible to sort the information by the states of any particular descriptor if there is a lexicon available. For names, whether of species or of persons, e.g. collectors, and for numbers, this presents no difficulty, and there is probably no necessity to compile a lexicon from the data bank. If any sorting by locality is to be carried out, as it clearly will be, some standardised system of recording localities will need to be adopted, and such systems have been developed. No comprehensive system for recording habitat data has, however, been proposed, and the compilation of one into which entries on herbarium labels can be translated will clearly be very difficult, if not impossible. Nevertheless, extraction of data from a bank by habitat will only be possible if some such system is designed.

In designing the data bank for any collection it will be necessary to consider for each of the descriptors to be adopted whether it will be desirable to sort the data bank by the states of that descriptor, and, if so whether the entries under that descriptor can have a form that will make this possible.

To those already acquainted with the data-processing systems already suggested for handling data about taxonomic collections, little, if any, of what is said above will be original. My excuse for putting it forward is that, together with the two papers that precede it, it will, if we have succeeded in what we set out to do, make plain the requirements that any system for the automatic processing of data about herbaria and other taxonomic collections must meet.

ELECTRONIC DATA PROCESSING OF HERBARIUM SPECIMENS
DATA FOR THE FLORA OF VERACRUZ PROGRAM*

A. Gómez-Pompa, J. A. Toledo and M. Soto

Department of Botany, Institute of Biology

National University of Mexico

Summary

The Flora of Veracruz program is a comprehensive study of the plant resources of this Mexican state. It involves a series of studies that go beyond the common objectives of the classical floras as it includes ecological and environmental studies and it is utilizing electronic data processing methods for the entire project.

The main objective is the production of a Flora of the same type as an ordinary good modern one. At the same time it could serve as a basis to create a data bank on the plants and the environments of Veracruz which could have many uses throughout the time of preparation of the Flora and afterwards.

A brief discussion of the kinds of data that are being introduced in the data bank will be explained here. We will present a more ample discussion of the data from herbarium specimens.

* Flora of Veracruz, contribution number 15. A joint project of the Institute of Biology of the National University of Mexico and the Department of Botany of the Field Museum of Natural History, Chicago, U. S. A. to prepare an ecological and floristic study of the state of Veracruz, Mexico.

Information about this project was published in <u>Anal. Inst. Biol. Univ. Nat. Méx. Ser. Bot.</u> 41: 1-2. Partially supported by N.S.F. grant GB-20267X.

Even though most of the effort has been concentrated on new collections, we have made a survey of the collections already present in the herbaria of several institutions. As an example of this effort we will present the pilot project that has been accomplished to compile the data from Veracruz specimens from the herbarium of the Royal Botanic Gardens at Kew. We will show examples of the kind of processing that we are doing with them.

A comparison of the collections of two families (Piperaceae & Araliaceae) in selected European herbaria will be presented as another example of the kinds and numbers of Veracruz collections housed in these herbaria.

A discussion on the future of herbarium data processing in future botanical research will be presented in the context of our floristic research programme.

Introduction

Herbarium specimens are one of the most important sources of information for any floristic project. They provide information about historical collections of great value for nomenclatural purposes, as well as information about areas where today we find only agricultural land or cities. Herbarium specimens also provide geographical data that are basic for protection and conservation of rare species. They supply unique information that cannot be found anywhere else about the distributions of all taxa. One of the greatest values of old collections from tropical areas is that many of them may be the only source of information about species that today may be extinct. This may be especially true for collections that came from areas now densely populated in tropical evergreen rain forest regions. There are other types of data available in herbaria that may be of interest, such as local names, plant uses and ecological information.

To capture these data the Flora of Veracruz Program has developed an information system for processing various types of data concerning the environments and plant resources of Veracruz.

One of the great problems for any newly initiated floristic study is to establish the quantity and quality of collections previously obtained in the area under investigation. This difficulty arises because collections are scattered in many different herbaria throughout the world, and herbaria are not organized in such a way that the information can be easily obtained. Another important problem is the change in scientific names caused by revisionary and monographic work and it is also for this reason that a dynamic information system is needed.

For these and other reasons (Gomez-Pompa & Nevling, 1973) the Flora of Veracruz Program was planned from the beginning to use electronic data processing methods. One compelling reason for employing the computer in this Program was the need to have a Flora that could be useful even during its development. Normally a Flora is useful only after it has been published, because no information of a general nature can be obtained until it is completed. Another important reason for E.D.P. is that it should facilitate the rapid production of related Floras, of both more and less inclusive areas, once the initial Flora is completed. Classical Floras do not provide ways by which the information gathered for one Flora can be used for the next (Shetler 1971).

Among the sources of data for this Program, label data from herbarium specimens has had high priority. Three years ago several projects were started to capture data from specimens in selected large herbaria with the objective of evaluating the quality and quantity of Veracruz collections in those herbaria. At the same time we have captured the label data for the Flora of Veracruz data bank.

We have surveyed the following herbaria for this purpose: Arnold Arboretum, Gray, the Royal Botanic Gardens, Kew and the National Herbarium of the University of Mexico. In addition, we have made a sample survey of two families (Araliaceae & Piperaceae) in a larger number of herbaria: Leiden, Utrecht, Paris, Florence, Vienna Museum, British Museum, (Natural History), Missouri Botanical Garden, Field Museum, and Madrid.

We thought that by approaching the problem in this way we could have a good idea of what collections from Veracruz are available in the herbaria of the world. This paper presents some of the results obtained from the European collections, that will help us to explain this part of our E.D.P. system for the Flora of Veracruz.

Methods

In each herbarium in which a search was conducted the first step was to locate the specimens from Veracruz. The most serious problem encountered was with those specimens that do not have a clear indication that they come from Veracruz. In these cases the only clue was the locality name, or a combination of the collector's name and locality. All doubtful cases were recorded for future checking.

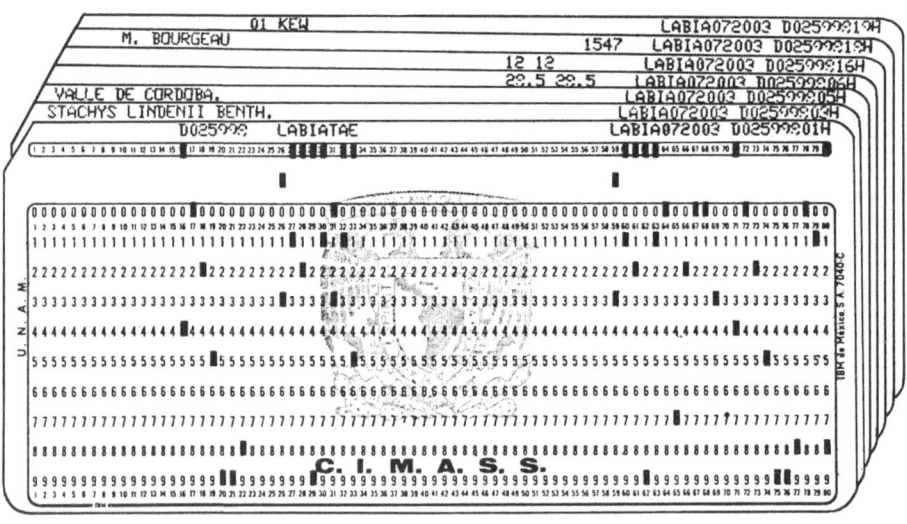

Figure 1.   Pre-printed form used for capturing the data from herbarium
specimens and its corresponding set of punched cards

When a Veracruz specimen was found, a standard method designed to capture data was followed (Gomez-Pompa & Nevling, 1973). This consisted of filling out a pre-printed form, with a fixed format, to facilitate the preparation of punched cards for input into the data bank (Fig. 1). This process was followed with all specimens found; the only exceptions were specimens in type folders. In this case a special effort was made in some cases to verify from the literature if the specimen was in fact a type, and then a note that the type had been checked was included on the form.

The completed forms were sent to the computer laboratory of the Flora of Veracruz where the data were reviewed and prepared for entering into the data bank.

Two types of information from the labels are recognized: the first is the precise information concerned with the scientific name of the plant, the locality, and the collector's name, number and the date; the second type includes ecological information, plant uses, local names, etc. The first group is treated very rigidly for editing purposes and retrieval, whilst the second type is mainly stored in full text with very flexible format, thus allowing retrieval without standardisation of the information stored under these categories. This division of information permits us to capture all the information available on the labels without losing time in checking the reliability of the information, although this may be done later if desired.

The computerized system for the Flora of Veracruz includes an automatic method for checking and detecting errors in the information to be entered in the data bank, especially errors in the first type of information mentioned. To facilitate this process certain elements were codified, viz. scientific name and locality.

This codification of the scientific names is done using a basic master file of names in the system. This file makes it possible to work automatically and dynamically with all scientific names known for Veracruz plants. The use of this file has proved to be useful in preventing nomina nuda from entering the data bank. When possible, the locality was coded according to our system, which is based upon a latitude and longitude grid. In many cases, however, coding could not be carried out because of vague locality data or because many old place names are no longer in use and need special research for locating them.

After the foregoing steps in data preparation are completed, a set of punched cards is produced, thus converting the data into computer-readable form.

This first information input is automatically checked by the system and it only accepts that information which is error-free (according to our master files). Incomplete or erroneous information is refused with its problem diagnosis message (Fig. 2).

We want to emphasize that the data obtained from herbarium specimens is not changed or lost in this process and that we consider the information as working data to be used and evaluated by the taxonomists who will write the family treatments. Such information alone is not regarded as an end product.

The method followed in the Flora of Veracruz Program is a simple one that allows the input of information in the easiest way for future retrieval.

Once in the computer the data can be processed in many different ways. We can have a full listing of all the collections in one herbarium (Fig. 3), a list of the collections of one collector in one family (Fig. 4), a list of the localities of one collector (Fig. 5), statistical data of one collector (Fig. 6), or a list of type specimens (Fig. 7). It is important to note in relation to type specimens that a major problem results because not all specimens annotated as types are indeed types.

There are several other combinations that can be made with the information of the data bank. Figure 8 illustrates some statistical data relating to two families, and Figure 9 shows part of a chronological list arranged by collectors.

Discussion

The number of collections (2879) from Veracruz in the families surveyed up to now at Kew was lower than we optimistically estimated in the initial planning of the survey. Because of this difference a spot-checking survey was started in order to evaluate the thoroughness of the general survey. Even though the total number of specimens may be low, the value of the Kew holdings is great because of their high quality and historical importance. But the most significant aspect of this investigation in European herbaria is that we now have another basis for planning field work and herbarium studies with old collections.

Concerning the holdings of other herbaria, even though the number of taxa sampled may not be statistically representative, they can give us a clue as to total holdings by indicating the collectors represented and their

```
27363 NOMBRE DE ESPECIE INCORRECTO    RENEALMIA MEXICANA KLOTZSCH EX PETERSEN
ZINGIBERACEAE      RENEALMIA
27363 COLECTOR AUSENTE O INCORRECTO
27363 NUMERO YA EXISTENTE
27364 NUMERO YA EXISTENTE
27185 MAPA AUSENTE O INCORRECTO
27185 FECHA AUSENTE O INCORRECTA
27185 NUMERO YA EXISTENTE
27187 TARJETA 1 O 3 AUSENTE
27187 COLECTOR AUSENTE O INCORRECTO
27187 HERBARIO AUSENTE O INCORRECTO
27187 NUMERO YA EXISTENTE
27367 NUMERO YA EXISTENTE
27368 NUMERO YA EXISTENTE
27369 NUMERO YA EXISTENTE
27371 NUMERO YA EXISTENTE
23662 NOMBRE DE ESPECIE INCORRECTO    TECOMA STANS (L.) HBK.
BIGNONIACEAE       TECOMA
23662 NUMERO YA EXISTENTE
27109 NOMBRE DE ESPECIE INCORRECTO    ANTHURIUM SCANDENS VAR. VIOLACEUM SCHOTT
ARACEAE            ANTHURIUM
```

Figure 2.   A set of diagnosis errors sent by the computer

```
025864    GUTTIFERAE
          CLUSIA MEXICANA VESQUE.
          28.5/28.5 VALLE DE CORDOBA.
          M. BOURGEAU                1957        17/02/
          HERBARIO K


025865    GUTTIFERAE
          CLUSIA ORIZABAE HEMSL.
          28.5/25.0 IZHUATLANCILLO ORIZABA.
          M. BOURGEAU                2967        14/08/1866
          HERBARIO K        *TIPO*


025866    GUTTIFERAE
          CLUSIA ORIZABAE HEMSL.
          28.5/28.5 VALLE DE CORDOBA.
          M. BOURGEAU                1954        16/03/
          HERBARIO K


025867    GUTTIFERAE
          CLUSIA PARVICAPSULA VESQUE.
          28.5/29.5 POTRERO PRES CORDOBA.
          M. BOURGEAU           S.N.         14/01/
          HERBARIO K


025868    GUTTIFERAE
          HYPERICUM PHILONOTIS CHAM. & SCHLECHT.
          29.5/24.0 LOMA GRANDE MT. ORIZABA.
          E. K. BALLS           5380        27/08/1938
          HERBARIO K


025869    GUTTIFERAE
          HYPERICUM OAXACUM R. KELLER.
          40.0/25.5 BELOW LAS VIGAS PEROTE.
          E. K. BALLS                4790          /  /
          HERBARIO K


025870    GUTTIFERAE
          HYPERICUM OAXACUM R. KELLER.
          29.5/24.0 LOMA GRANDE ORIZABA.
          E. K. BALLS                5364        27/08/1938
          HERBARIO K


025871    GUTTIFERAE
          HYPERICUM CONFUSUM ROSE
          37.5/25.0 ACTOPAN, COFRE DE PEROTE.
          E. K. BALLS                5553        12/09/1938
          HERBARIO K
```

Figure 3.   Portion of the list of collections of Veracruz plants at
             Kew, obtained from the data bank of the Flora of Veracruz

```
BROTEN.
   BROMELIACEAE
          1052   027085
COSSON E.
   ARACEAE
           409   027186
   GRAMINEAE
            48   025701
COULTER DR.
   ACANTHACEAE
          1209   023673
   ASCLEPIADACEAE
           979   023631
           980   023618
   BORAGINACEAE
          1062   023696
   CARYOPHYLLACEAE
             -   023741
           694   023719
   COMPOSITAE
           370   024069
           392   023764
           400   024053
           436   023899
           452   024166
   CONVOLVULACEAE
          1012   024510
   GRAMINEAE
          1643   025614
   LILIACEAE
          1590   027401
   MELASTOMATACEAE
```

Figure 4.   Fragment of a list of collectors of Veracruz collections
            at Kew extracted from the data bank of the Flora of Veracruz
            which includes: names of collectors, family, collectors
            numbers and accession numbers to the data bank

```
      2617 REGION DE ORIZABA, SAN CRISTOBAL.
   32.5/31.0
      10687 ZACUAPAN
   33.0/40.0
        618 VERACRUZ.
   33.5/29.0
        182 MIRADOR.
       4281 CORDILLERA VERACRUZ
   43.5/33.0
       2968 SANTA TERESA, REGION DE ORIZABA.
   44.5/29.5
       2452 MISANTLA
BROTEN.
   28.0/26.0
       1052 ORIZABA.
COSSON E.
   28.0/26.0
         48 ORIZABA.
        409 ORIZABA.
COULTER DR.
       1209 SIN ONLY VER.
   33.0/40.0
          - VERACRUZ
   38.0/28.5
          - JALAPA.
          7 JALAPA
        125 JALAPA
        129 JALAPA
        133 JALAPA.
        370 JALAPA.
        392 JALAPA.
        400 JALAPA.
        436 JALAPA.
        452 JALAPA.
        694 JALAPA.
        773 XALAPA
        979 XALAPA.
        980 JALAPA.
       1012 JALAPA.
       1062 JALAPA.
       1193 JALAPA
       1238 JALAPA
       1299 XALAPA
       1397 JALAPA
       1590 JALAPA.
       1643 JALAPA.
DE LEON CARLOS
   32.5/28.0
          - HUATUSCO. VER.
FINCH HUGO
   28.5/28.5
```

Figure 5.   Partial list of collectors from specimens at Kew that
             includes: collection numbers, localities, locality codes
             and names of the collectors

	K	P	L	FI	BM	WV	Total
Number of Specimens	572	71	20	12	4	1	680
Number of Collections	542	54	20	12	4	1	633
Number of Duplicates	53	28	17	12	4	-	161
Number of Unicates	489	26	3	-	-	1	
Number of Families	64	2	2	2	2	1	65
Number of Genera	211	4	4	2	3	1	211
Number of Species	408	38	13	7	4	1	408

Fig. 6a

Figure 6, a & b.   Samples of statistical data of part of the collections
of M. Bourgeau.

Collection number	K	P	L	BM	FI	Total
1491	1	1	1		1	4
1492		2	1		1	3
1493	1	1			1	4
1512	2					2
1559	2					2
1569	2					2
1606		3	1			4
1677		1	1			2
1759	2					2
1804		6				6
1805	1	1				2
1806		2				2
1847	2					2
1853	2					2
1866		1		1		2
1958	1	1	1			3
1959	1	2	1		1	5
1988	1	1	1			3
1989	1	1				2
2054	1	2	1		1	5
2168	1	1	1		1	4

Fig. 6b

```
024576   ACANTHACEAE
         JACOBINIA INCANA HEMSL.
         33.0/40.0 VERACRUZ.
         J. LINDEN                1079          /  /1838
         HERBARIO K      *TIPO*

025574   GRAMINEAE
         COELORACHIS RAMOSA (FOURN.) NASH
         27.5/25.0 RIO BLANCO PPCES ORIZABA.
         M. BOURGEAU              2647      24/07/
         HERBARIO K      *TIPO*

025603   GRAMINEAE
         PASPALUM MINUS FOURN.
         28.5/28.5 VALLE DE CORDOBA.
         M. BOURGEAU              2298      24/04/
         HERBARIO K      *TIPO*

025613   GRAMINEAE
         PASPALUM SQUAMULATUM FOURN.
         27.5/25.0 RIO BLANCO PRES ORIZABA.
         M. BOURGEAU              2640      24/07/
         HERBARIO K      *TIPO*

025660   GRAMINEAE
         PANICUM VISCIDELLUM SCRIBN.
         38.0/28.5 GRAVELLY BANKS NEAR JALAPA.
         C. G. PRINGLE           8089      29/03/1899
         HERBARIO K      *TIPO*

025717   GRAMINEAE
         MUEHLENBERGIA SETARIOIDES FOURN.
         28.0/25.5 ORIZABA. BORREGO.
         M. BOURGEAU             3362      14/11/
         HERBARIO K      *TIPO*

025738   GRAMINEAE
         EPICAMPES BOURGAEI FOURN.
         28.0/26.5 ORIZABA. ESCAMELLA.
         M. BOURGEAU             2973      26/08/
         HERBARIO K      *TIPO*

025787   GRAMINEAE
         ELEUSINE INDICA (L.) GAERTN.
             /       VERACRUZ.
         DR. GOVIN               S.N.          /  /1867
         HERBARIO K      *TIPO*

025819   GRAMINEAE
         ARUNDINARIA ACUMINATA MUNRO
             /       VERACRUZ.
         LIEBMAN.                  73          /  /
         HERBARIO K      *TIPO*
```

Figure 7.   Portion of a list of the types from Veracruz found at
            Kew and entered in the data bank.

Familia Araliaceae

	K	L	P	W	WU	BM	FI	Total
Number of Specimens	20	3	14	7	1	11	4	60
Number of Collectors	8	2	6	5	1	5	2	11
Number of Collections	20	3	14	7	1	11	4	60
Number of Genera	3	2	2	3	1	2	1	4
Number of Species	5	3	4	4	1	3	1	7

Familia Piperaceae

	K	L	P	U	WU	BM	FI	Total
Number of Specimens	84	40	229	36	8	59	23	479
Number of Collectors	16	7	23	4	2	15	7	34
Numbers of Collections	84	40	205	35	8	59	23	554
Number of Genera	2	3	3	3	2	2	2	3
Number of Species	36	23	57	25	6	26	9	118

Figure 8.   Statistical data of the collections of Piperaceae and
             Araliaceae from the survey made in several European
             herbaria.

```
GOULD  FRANK  W.
     17  09  1965    11710    028132
GOVIN.
              1865        2    027891
              1866        7    027878
              1866        8    027871
              1866        9    027870
          01  1866        6    028112
     28   01  1866       11    028114
     11   08  1866        3    027919
              1867        -    028060
              1867        -    028076
              1867        -    027729
              1867        -    028131
              1867        -    027863
              1867        -    026793
              1867        -    025787
              1867        1    027892
              1867       10    027931
              1867       12    027954
              1868        4    028125
              1868        5    028126
              1869        9    028127
GRAY  A.
          05  1885        -    024077
GREENMAN   J.  M.
     24   06  1906      116    027379
HAHN  M.
                         -    023629
                         -    026570
          03             -    023599
     30   06             -    026607
          08             -    026745
          09             -    023935
              1865        -    026634
              1865        -    026639
     13   04  1865       46    028120
     11   08  1865        -    027872
     12   08  1865        -    027966
              1866        -    028118
              1866        -    027774
              1866        -    027932
              1866        -    024532
     16   03  1866     2066    028080
     16   03  1866     2066    027768
     18   04  1866       46    026805
     18   04  1866       59    028122
     25   05  1866        -    027216
     25   05  1866      189    027977
```

Figure 9.   Portion of a list of collections arranged by collectors and
            also chronologically from the holdings at Kew and stored
            in the data bank.

collection numbers. It was evident that the herbarium at the Museum of Natural History at Paris is of high priority for our research.

One problem in maintaining an up-to-date file is that any change in the determinations of Veracruz specimens has to be communicated to us. We trust that most of the corrections will be made through specialists working for our programme. If this is carried out the problem of updating does not exist. Another possibility is that when monographers change the names of specimens, we wait until the paper (monograph or revision) is published to capture the corrections. When a curator notes a mis-determination and routinely changes the name, it will be almost impossible for us to enter this change in our files. For this reason we would greatly appreciate having sent to us information regarding any changes in the names of Veracruz specimens. Although we recognize the difficulties involved in this request, we hope that the problem will eventually be solved by common agreement among curators. However, our experience is that the name of a plant from Veracruz has little like-lihood of being changed except through our own work or by monographers, so the risk of losing information is minimal.

We have demonstrated to our satisfaction that the general system designed for capturing label data for Veracruz specimens applies equally well to old herbarium specimens and new material, and the system is operational. It is important to mention that the most significant limiting factor is the personnel doing the data capture. They have to have a certain degree of taxonomic knowledge and a great amount of patience to do routine work with responsibility. It is our experience that some training and revision has to be incorporated in any survey in order to be sure of the correctness of the information gathered.

Acknowledgments

Data for this paper were obtained from the data bank of the Flora of Veracruz, those from Kew were obtained by Mrs. Laura Vit and those from other European herbaria by the senior author with the assistance of Mrs. Norma Gómez-Pompa. We wish to express our gratitude to these persons and the curators of the herbaria surveyed for their help. Also we want to thank Dr. John Beaman for his comments and corrections to the original manuscript.

REFERENCES

Gómez-Pompa, A. & Nevling L. I. Jr. (1973). The use of
        electronic data processing methods in the Flora of
        Veracruz Program. Contr. Gray Herb. 203: 49-64.

Shetler, S.G. (1971) Flora North America as an information system.
        BioScience 21:524-532.

Vit, L. The collections of Veracruz at the Royal Botanic Gardens
        at Kew. Thesis, Facultad de Ciencias, UNAM, México
        (in preparation).

THE USES OF ELECTRONIC DATA PROCESSING FOR HERBARIUM SPECIMEN LABEL INFORMATION

A. V. Hall

Bolus Herbarium

University of Cape Town, South Africa

Summary

An introduction is given to the chief elements of computer-based information systems for data on herbarium specimen labels. Some critical reasons for considering using such systems are noted, particularly for providing multi-indexed access to label records. It is concluded that with careful systems analysis, the computer-based approach should be of major value in extending the usefulness of the herbarium.

Introduction

Computers have been proposed as aids to the taxonomist in a variety of ways. These range from the simplest tasks to the most complex, such as:

1. printing determinavit labels and distribution maps

2. information management (storage and retrieval)

3. automatic key-forming and

4. forming groups of taxa and showing their cores, edges and other internal structures, as well as showing trends.

The last two of these groups of procedures need a numerical image of the taxa, so that the work has been known as 'numerical taxonomy', or better as: 'numerical aids to classification'. The use of these numerical aids has been hampered by problems in treating the data in a manner that is satisfactory to taxonomists. Only rather recently has there been an emergence of systems that meet their demands in a reasonably rigorous way (Roth, 1970; Hall, 1973).

By comparison, using the computer to assist with managing specimen label information is a much simpler process. There are three basic elements:

1.   Data preparation: Specimen label information is typed into a card or paper-tape punch, or into a magnetic tape encoder.

2.   Forming a data-bank: The label information in its punched card or tape form is transferred into the computer, with some automatic checking, provision for proof-reading and arrangement for correction. The compiled information forming the data bank is entered so as to be as accurate as possible. The likelihood of incorrect taxonomic naming is provisionally accepted.

3.   Obtaining results from the data-bank: Detailed requirements are given to the computer for the kind of information needed. The results are given after automatic searching and sequencing has been carried out. The results may be printed, plotted or displayed on a screen.

Data-Banking Procedures

An introduction to the main features and problems in data banking is given in the following sections. The introduction is based on experience with a small system that is being used as a monographer's aid, and on studies of large-scale approaches in use at a variety of institutions with major collections (Crovello & MacDonald, 1970; Hall, 1974).

1.   Data: The label information, in the state we find it on the herbarium sheets, is the raw data for the processing system. It may have serious imperfections such as unclear handwriting, missing or inferred data such as collector or altitude, together with problems such as obscure, out-of-date or vague place names. The raw data may have to be interpreted, re-arranged and supplemented with standard codes such as latitude and longitude values. The changes are made so

that the data may meet certain standards for efficient processing. In some systems, abbreviation codes are used for lengthy statements that appear frequently.

2.   Input : The term input is used both for the set of data being entered, and for the actual process of putting the information into the computer. The input often takes the form of cards, punched with the information on a typewriter-punch that gives a line of printed data along the top of the card and holes coded in a pattern for each respective letter or number, below. The cards are passed through an automatic reader and the hole-patterns are converted into the computer's internal symbols of short strings of on/off memory stores.

3.   Programs: A program controls the movement and processing of information in the computer. It consists of a set of instructions written according to standard rules. The rules are set so that the instructions can be 'understood' by a highly 'intelligent' program in the computer called a compiler. A set of instructions for use with a compiler is known as a program language. The chief general languages are known by their acronyms, which reflect the chief uses for which they were originally intended:

> FORTRAN (formula translation)
> ALGOL   (algebraically-orientated language)
> COBOL   (common business-orientated language)

Compilers frequently have small, but sometimes significant, differences between one computer and another. Some languages have gone through several generations of development, so that there are, for example, three frequently-used versions of FORTRAN, quoted as II, IV and V. FORTRAN IV is perhaps the most widely used for general scientific purposes. Some workers have written their own special-purpose compilers and languages. These are generally difficult to move from one computer to another. Larger computers will certainly have compilers for the chief languages. In considering adopting a data banking program, it is important to know that its language can be interpreted efficiently by a compiler in the computer to be used, especially if it is a small-capacity machine.

4.   Editing, proof reading and correcting: Some computer-based editing of the input is usually possible, but the major part of the checking seems best carried out by reading computer-printed proofs of the input to another person, who checks the data with the original labels. This is an essential but tedious chore, exemplifying the major effort that is needed in building up the data bank. Many errors appear in the repetitive typing of unfamiliar collectors' names, and there is a good case for being able to use abbreviations that can be programmed to call out the full spelling from a 'directory'. The names then need only be entered once into the computer, and after checking, can appear correctly for hundreds of specimens. A wrongly spelt or numbered abbreviation will mostly give a glaringly wrong result that will be noted during proof reading. When errors have been found, a simple means of entering corrections into the data bank is most important.

5.   Data-bank files: Once the input has been completed, the information will be stored in one or more 'files', which are made up of very large numbers of sequenced storage positions in the computer. The files make up a data-bank. A record in a file may be, for example, the data from a specimen-label. The records may be linked by indexing into groups and super-groups, giving a hierarchy. The structure of hierarchies can be highly significant in retrieval problems. It is important to know that a program that may be considered for data-banking can manage hierarchies in an effective way.

6.   Retrieval: The process of finding some set of information in the data-bank is known as data-retrieval. For many queries for small data-banks, retrieval may be very simply carried out, using bound catalogues, with the contents sequenced and printed in various ways by the computer. Such an approach may be quite efficient enough in many working environments. In other cases, retrieval may have to be carried out under program control, using an input pattern-template showing the system what kind of record is wanted. Retrievals may be made to answer questions such as the following:

    a.   What known types of Genus A exist in the herbaria
          recorded in the data bank?

    b.   What species have been collected at the higher altitudes
          on a certain mountain range in September and October?

    c.   What species were recorded in an area prior to a period
          of drastic environmental change?

    d.   What was the range of localities visited by a collector
        in a particular year (as an aid in finding an obscure
        type locality)?

   7.   Sequencing and tallying: One or more criteria may be used to
set the records in sequence, such as by month of collection, by
altitude or by latitude. The success of this will depend on the
completeness of the label data. Tallying is the counting of the number
of cases for various versions of a kind of datum, among the retrieved
records. Where the data permit, useful tallyings may be the
frequencies of a taxon at various altitudes, or the numbers of
collections of flowering material of a species in each month of the year.

   8.   Mapping: Maps showing geographical distributions can be drawn
automatically if latitude/longitude values have been included in the data
bank. While the maps can be extremely useful, it is noted that the
finding and checking of the values may be one of the most time-consuming
aspects of data preparation. Checking is assisted if the names of major
geographical features are assigned to each grid-square, with the name
being automatically called for inclusion in the listing of each record for
proof reading.

   9.   Special features: The above descriptions give the bare essentials
of a data-banking system. Refinements can be added, especially for
easing the major burdens of data preparation and input, as well as
providing for quick retrieval and sorting. Important provisions should
be made for keeping the user's instructions very simple so that he may be
able to follow a series of obvious steps on coming to use the system for
the first time, or after a lengthy absence. Extra programming may be
needed to achieve this simplicity but it certainly proves most worthwhile.

Discussion: The Uses Of Data-Banking.

With the above background of procedures, one may ask what general
improvements can be expected in return for the major effort that is
needed. Are the present arrangements for using herbarium-label
information good enough? The answer is that although the herbarium
curator may have enormous amounts of data close at hand, it cannot
readily be used for many purposes because it is usually only filed
in a taxonomic way. Ask the curator what is the high-altitude flora of
a particular region and it may take a good week or more of 'herbarium
crawl' techniques to get a set of results. In the past, the herbarium
curator probably couldn't answer the problem well because there were

too few records. Today, the curator cannot because there are too many.
In this respect the herbarium could perhaps be seen as less of a data
bank than a 'data crypt': because of no multi-indexing one must look
for long hours for facts buried in a mass of irrelevant material.

In fact, one seldom attempts problems that need large-scale multi-
indexing: they are too time-consuming. A computer-based information-
management system changes this position in a quite dramatic way.
Indeed, it improves the usefulness of the herbarium collection to a
rather important degree, especially for fields such as biogeography.
It allows one to reap the full benefit of perhaps thousands of man-years
of field recording and taxonomic study, much of which may appear
only in summary form in regional floras in many parts of the world.
Full and precise notes for labels would be seen to be more widely
useful than at present, providing a valuable incentive to giving better
field data.

In many cases, the curator would be aided by the data banking system
in controlling the growth of his collection. With the aid of geographical
grid data, the more thinly represented regions could be shown.
Indications could be given of under-collected taxa. The curator could
give his collectors check-lists of the taxa already known from areas
being re-explored. Experience has shown the great value of this in
building up part of a regional herbarium of the University of Cape Town.
Wasteful repeated collection of the same taxon at the same place is
avoided.

If the procedures for information management can be kept simple,
they can be a useful aid to the monographer in organising his results.
Clearly printed listings from other herbaria would help the monographer
interpret the obscurities of labels on borrowed specimens. The correct
interpretation of poor handwriting or out-of-date place names may be
relatively easy for the curator of the specimens but most difficult for
a borrower, sometimes leading to mis-interpretations and loss of
information. Important aids could be given to finding type specimens.
A register of types could be extracted from a specimen-label data
bank, for the information of workers within the herbarium and elsewhere.
Where the types are not labelled as such, the monographer may want to
see listings of label data for all specimens at a herbarium that belong to
his group. One would then have the option of borrowing only the most
needed material, reducing the size of at least some loans and perhaps
lessening the need to visit remote herbaria. The practical effect of this
in speeding up revision work would certainly seem highly desirable.

On another aspect, experience suggests that the work of attracting and training new recruits to taxonomy may well be eased where data-banking aids are available.

Conclusions

It would seem that data banking can certainly offer immediate benefits, especially to monographers and the curators of regional herbaria. Data banking for all specimens in the major herbaria would provide valuable long-term assistance to taxonomists, apparently every bit as important as the local or regional projects. However, for very large collections there is the acute practical problem of the major effort needed to get the information into the computer. A wide variety of problems exists in proposing data banking in the special conditions in which each herbarium finds itself. In fact, careful individual studies of priorities need to be carried out before work on data banking is started. Known as systems analysis, such a study involves a thorough examination of available equipment and programs, fundings, personnel, the amount and nature of the data, and the likely usage of the data bank. A study of this kind is the first step in adopting procedures that appear to have important advantages for the users of herbaria.

REFERENCES

Crovello, T. J. & MacDonald R. D.(1970). Index of EDP-IR projects in systematics. Taxon, 19: 63-76.

Hall, A. V. (1973). The use of a computer-based system of aids for classification. Contrib. Bolus Herbarium, 6: 1-110.

Hall, A. V. (1974). Museum specimen record data storage and retrieval. Taxon, 23 : 23-28

Roth, H. D. (1970). Cluster analysis for the biological and social sciences. Smithsonian Inst. Systems Innovations, 2 (2): 1-35.

DISCUSSION, AFTERNOON, WEDNESDAY, 3 OCTOBER

This discussion, led by Dr. R. M. Cowan, was a general one on the papers read during the afternoon, and additionally on those given in the morning.

Dr. R. M. Cowan (Introduction). I am here acting as the representative of the International Association of Plant Taxonomy and convey greetings from Dr. F. Stafleu to this Conference. I. A. P. T. is willing to give any help it can to the establishment of an information management system to serve taxonomy. It is hoped that a computer print-out of Index Nominum Genericorum will be ready for the Botanical Congress in Leningrad.

Mr. R. J. Pankhurst. Standards are necessary for the identification of specimens especially the descriptive data. These standards also apply to non-taxonomists.

We must become accustomed to having our data as complete and consistent as possible. This is absolutely necessary for computerisation of such information. With this data it should be possible to evaluate automatic identification on a large scale. It is vital that a national institution should be responsible for the task.

Professor J. Heslop-Harrison. One of the most interesting details is the cost. How did Professor Gómez-Pompa's Veracruz project come into being?

Professor A. Gomez-Pompa. The Flora of Veracruz Project started in 1966. The idea for a Flora of Veracruz and the use of E. D. P. as its base came at the same time. A new project for the National Herbarium of Mexico was needed in order to increase the interest of the National University of Mexico and to encourage new botanists in the area of floristic botany. A pilot project to computerize the Herbarium was started with the help of the Computer Center of the University. Also a pilot program for using E. D. P. methods in the projected Flora of Veracruz was tried. Having succeeded with these projects, contacts were established to determine who was working on similar projects and an International Symposium on Information Problems in Natural Sciences was organized by our University with the co-operation of the Smithsonian Institution. In that Symposium the initial results of our two pilot projects were presented. At the same time, the Flora of Veracruz project started to be a reality and I decided to associate the Flora with an institution having a large herbarium of Veracruz plants, good library facilities for students and personnel willing to embark on a joint project. These facilities were offered by Harvard University and since then a co-operative scientific venture was initiated with the collaboration of Dr. Lorin I. Nevling Jr. In the recent past, and because of the change of Dr. Nevling to Chicago, the project is now between the National University of Mexico and the Field Museum of Chicago. The student collaboration has been kept with Harvard. The initial funds for the project came from the National University of Mexico and from Harvard University. As the scheme of the Flora has been succesful in many ways interest for the project grew and additional funds became available.

Professor J. Heslop-Harrison. What was the level of skill necessary for an operative to extract data from the labels?

Professor A. Gómez-Pompa. Technicians who were biology undergraduates were used in gathering the data from the herbaria of Kew, Harvard and Mexico.

Dr. A. Hall. Technical assistants did this work in the Bolus Herbarium

Mr. J. Raynal. How many specimens were dealt with and how were the names checked?

Professor A. Gómez-Pompa. About 30, 000 specimens were examined. All the identification of new material was done or checked by myself at Mexico, and by Lorin Nevling at Harvard (now at the Field Museum, Chicago) and other specialists who co-operated in the programme. The older material was not checked but corrections can be made when monographers have worked on this material.

Dr. D. J. Rogers. What kind of programmes were used and are these available?

Dr. J. A. Toledo. We use our own system, written in ALGOL language. The system is not available for general study.

Dr. R. M. Cowan. It is necessary that the data should be of value to other users in order that granting agencies will advance funds.

Professor A. Gómez-Pompa. The ecological part of the Flora of Veracruz Program includes the gathering of geographic and environmental information. The objective is to understand the factors involved in the distribution of taxa and to understand the ecology of the vegetation and its evolution. For example, all the available climatic information has been included in our data bank as well as the geographical information. This means that several hundred thousand items are available for E.D.P. This information has proven to be of great value to several unrelated governmental and private agencies.

Dr. J.F. Mello. We are in the process of creating over 50 separate data files at the National Museum of Natural History in Washington, using the SELGEM system. For these 50 files the average cost for entering data into the computer system is $1.80 per specimen. Of this sum $1.57 is paid in salaries to employees who enter the data and administer the files, and 23c. is spent on computer time.

Dr. D.J. Rogers. What was the level of technical skill required?

Dr. J.F. Mello. A clerk/typist (Salary $7000 a year).

Mr. T.W. Davis. What machine was used in the Flora of Veracruz Project?

Professor A. Gómez-Pompa. A Burroughs 6700.

Mr. J.P.M. Brenan. Was there a significant amount of editing and how much of this E.D.P. exercise could one assess as part of the normal curation operations?

Dr. J.F. Mello. Yes, there was a significant amount of editing, but through the use of the computer to permute the files so that inverted listings could be prepared, much of the editing can be carried out by clerical personnel. On the average, information on from 50 to 100 specimens per day is entered by each operator (best performance so far has been 100 records in one hour). It is necessary to produce lists, labels and file cards as part of the normal curation procedure, and the computer has shown itself to be highly useful in producing these efficiently and accurately.

Dr. D.J. Rogers. How many items of data per specimen were recorded?

Dr. J.F. Mello. About 20.

Dr. J.L. Cutbill. On comparing costs at Cambridge it was 66.6p per item using the existing system and 65.6p per item by computer. Using a commercial quotation for computer time, the actual computing cost was a very small part of the total cost - about 1p per item. The question we should ask ourselves is: can we afford to document our collections properly anyway?

Dr. S.W. Greene. A part-time typist was used for the British Antarctic Survey Scheme. The running costs are small - £200 per year for machine maintenance and £150 per year for computing. We are able to put in as much information as we wish once per month and can question the computer once per month. About 5 days per month are taken in dealing with input, the rest of the time is used for standard curatorial duties.

Mr. R. Ross. Does the running cost quoted by Dr. Cutbill take account of programme writing?

Dr. J.L. Cutbill. No, it does not include such developmental costs. These were about £30,000 for programming costs. These are likely to increase to £100,000-150,000 for a fully operational system. The potential usage of the systems is an estimated half-million items being processed each year.

Mr. R. Ross. To what extent could we use an already developed system?

Dr. J.L. Cutbill. The system developed by the Museums Association (IRGMA) should be available.

Dr. S.G. Shetler. We should decide whether the goals we wish to achieve are valuable and if so to gather the necessary funds for carrying out an E.D.P. programme. Editorial manpower is necessary for publication of the data in order to ensure consistency. The figures which have been quoted do not tell the whole story.

Dr. R.S. Cowan. Many of these costs are add-on costs and must not be confused with the computing costs.

Dr. D.J. Rogers. It is wise to hire expertise to uncover costs which are unknown or overlooked. These management scientists can give details of costings and offer advice on ways of reducing them. An example of this was their suggestion of co-operation with other government agencies in the employment of disabled people who, with a short training programme, are able to carry out dull routine duties.

Professor A. Gomez-Pompa. In dealing with the problem of cost, we must ask first if the work to be done is worth it. If anyone embarks on an expensive E.D.P. project there should be a clear objective in mind. For example, the computerization of all the collections of the large European herbaria for the sake of it, is not worth it. But if the other objective is a World Flora the use of E.D.P. could be easily demonstrated. It seems to me that a World Checklist is a reasonable goal with E.D.P. methods, it would also have great practical importance. In a project with such an objective cost is not a relevant matter.

Professor C.H. Oppenheimer. We are required to go into a computerised system in order to handle the data efficiently. The urgent need is for compatibility to facilitate the interchange of information between the systems.

Dr. R.S. Cowan. The environmentalists must be made aware of the value of computerising this enormous amount of data. When they do, funds should become available. The question is, do we want the data which E.D.P. can provide for the use of the monographer?

Mr. J. Raynal. Compatible systems are not very important but it is essential that the information should be in a standardised format.

Mr. R. Ross. There is a curatorial need for a method which would enable us to find quickly all the material from a particular area; at the moment this is lacking. An agreed geographical breakdown, by means of which a list of geographic areas and their species could be made, is one of the priorities. The data capture for this is simple and the scheme worthwhile.

Professor J. Heslop-Harrison. I should query whether this is true. Most of the information required is to hand on opening the cover.

Dr. R.K. Brummitt. We are talking about two levels of information. Processes at the level of a taxon need have no relevance to the specimens, while there are other processes actually dealing with the specimens. In dealing with Mr. Brenan's list of priorities, nos. 1, 5 & 6 are specimen processes while 2, 3, 4, 7 & 8 are at the level of taxa.

I agree that we should concentrate on the problems at the taxon level. In a few years Kew's holdings could be dealt with in the way Mr. Ross suggests, which would be extremely valuable. It is important to deal with problems which have a definite end-point and which can be accomplished in a few years. Dealing with specimens is a very different matter. Even to cover the type holdings alone at Kew would be a very long term project.

Mr. P.S. Green. It might be possible to ask borrowers of material to extract the input data for E.D.P. and this could be supplied to an institution acting as a central European data bank.

THE FLORA NORTH AMERICA INFORMATION SYSTEM

S.G. Shetler

Flora North America

Washington, U.S.A.

The text of this paper is here given in abstract form only.

The Flora North America (FNA) Information System, as designed, consists of a precisely defined set of data files, collectively referred to as the FNA Data Bank, and equally well defined processing strategy and operating procedures based on the use of IBM's commercially marketed software known as the "Generalized Information System" (GIS). Each data file forms a module of the Data Bank and links by appropriate codes to related files. The files are designed to be used as parts of a whole or independently as authority files in their own right. The keystone of the Data Bank is the taxonomic name file to which every other file is referenced. During the initial development phase of FNA, the major file in terms of size and significance will be the one containing species-by-species summaries of morphological characteristics, habitats, and geographical distributions. The processing software, GIS, operates on System/360 and System/370 in batch mode under the supervision of IBM's standard teleprocessing monitors. GIS calls for formatted records defined in the system by data description tables and performs all of the basic functions of a generalized information-processing system, including report generation. It can interface with preprocessing and other special programmes. Two of its features of great significance to the FNA Information System are its capability of handling hierarchical data structures and its capability of processing multifile queries.

ELECTRONIC DATA PROCESSING IN THE HERBARIUM

F.H. Perring

Biological Records Centre

Monks Wood Experimental Station, Huntingdon, England

Summary

Experience of running a botanical data bank for nearly 20 years leads to the belief that the majority of questions can be answered if the data are available by species or by locality.

If records which carry data additional to locality are filed separately within the species file and all these data are supplied to any questioner, who can thus make his own selection of the data he requires, a two-way index is all that is necessary.

Such a simple system requires the minimum of E.D.P. and allows the majority of our slender resources to be used for collecting the data.

We question whether comprehensive E.D.P. provides the output which the user requires. The ideal for answering questions about species is the folder of specimens, but if that cannot be made available a photograph of the sheet or photocopy of the label contains much information, including handwriting, which is lost in a computer print out.

Microfilm and microfiche provide a system which ensures the safety of the record, is the basis of species and locality oriented information retrieval, and results in a format which can be sent economically by post throughout the world.

Introduction

The title given to this meeting 'The Use of Electronic Data
Processing Methods in Major European Taxonomic Collections' is one
which worries me considerably as it implies there are taxonomic
problems based on the use of museum material which can only be solved
by using a computer. The implication that computers are desirable and
essential in this and similar fields of biology is one which is very
frequently made, often I fear without questioning at the outset the
nature of the problem we have to solve, or asking whether there are not
simpler alternatives which would provide an adequate and cheaper
solution.

Dr. Shetler has indicated more elegantly than I ever can the problems
which can arise when applying E.D.P. to information retrieval in biology,
but, to my knowledge, Dr. Shetler's experience is not an isolated one.

It was in 1966 that the British Trust for Ornithology first decided to
use a computer to assist in the analysis of ringing recoveries, particularly
for common birds where returns which had accumulated over the last
30-40 years numbered several thousands. Now after 7 years, 3 computers
and 2 programming languages later, they are still without a satisfactory
programme, even though, like desperate men at the gambling table, they
have thrown more money into programming in the last 2 years in an
attempt to recover their losses. Fortunately for them they had a banker -
early on, to tide them over, they hired a counter sorter. It has been
producing results steadily since its installation, and several papers have
been published. This was essential in their case if their volunteers
were to continue to feel that the effort of ringing was worthwhile, and
to justify the costs of the project.

Even under my own roof at Monks Wood there have been problems.
About 8 years ago, in the early days of the International Biological
Programme, I was asked to assist in designing an information retrieval
system for the world series of nature reserves being assessed by the
CT (terrestrial conservation) section. Innocently I designed an 80-column
card system which I believed would be adequate to indicate which reserves
contained features of a particular configuration. I proposed sorting these
cards on a Pattern Select Sorter which would select cards carrying code
numbers to files containing information about the sites. However, about
1968 a meeting of influential scientists at Monks Wood decided that this

system was too unsophisticated: that it was incapable of answering all the questions they imagined might be asked, and I watched whilst the system became computerised. International Biological Programme money flowed like wine in a vintage summer. Consultants were engaged and enormous time and energy and £10,000 was spent in $2\frac{1}{2}$ years in developing a system which would, I am sure, answer any question you might care to ask. But supposing you don't care? At the end of 1972, when the Programme came to an end, very few other people did care, except the International Union for the Conservation of Nature who find the accumulated data store immensely valuable, would like to add to it, but can never afford the immense input and output costs of a computerised system, which in any case is far too sophisticated for their needs. It was therefore with a certain amount of amusement, and sorrow, that I learnt earlier this year that after another meeting of eminent scientists at Monks Wood the staff had been asked to develop an information retrieval system using 80-column cards. The questioner will receive copies of all the essential information about the sites matching his enquiry and will then use his own human computer to select certain features of particular interest; but this computer can not only select, it can judge the value of the data, consider alternative interpretations of unclear handwriting, and read maps.

A recent personal experience has emphasised to me the excellence of this human computer. For various reasons I wished to organise a symposium on Bracken (Pteridium aquilinum), not a plant I had previously studied, but I knew one person who had. A single phone call produced a list of 12 people currently working on the species. A dozen letters suggesting the possibility of a symposium and a request for names other than those I knew already, produced within 2 weeks what I am sure is a comprehensive list of research workers in both the pure and applied fields, and even included a list of names and addresses of scientists who had attended a one-day seminar on the subject 3 years ago who might be expected to become the core of an audience.

An alternative to this approach to an information retrieval problem in biology is the computerised titles only or enriched titles systems such as Medlars or Biological Abstracts Previews. During an experimental phase with the latter I was a customer with a British plant taxonomic profile. Every fortnight I received a sheaf of computer print out. Gradually the amount rose so that I now have several large boxes full of print outs - and I need an information retrieval system to work through them. I could have done this for Bracken, but it would have supplied me

only with recent references, not going back over 50 years - the service
supplied by my initial, octogenarian contact. He, by the way, because
he had all the reprints and had read them, was able to give me the names
and addresses of authors and some critical views on the value of their
work. These experiences lead me to believe that when concerned with
information retrieval we overlook the human computer at our peril, and
we may be in similar peril if we look uncritically at computers.
Computers really are splendid creatures for handling numbers: after
a relatively small amount of human effort in programming there are
immense rewards in time saving and accuracy, etc. - but they are
morons when it comes to information retrieval and an immense amount
of human effort in programming is required which is costly in time and
manpower. My view is that, in considering the problems facing the
major herbaria of Europe, we should use E. D. P. as little as possible.
This does not, however, exclude D. P. as I shall now try to explain.

Data Processing Of Individual Records At The Biological Records Centre

The Biological Records Centre holds over 2 million plant and animal
records and the data bank is growing at perhaps 100, 000 records per
annum. The majority of the records are simple distribution data
collected in the field and consisting only of a species number, a date,
a grid reference, a county number and a qualifier which may be status
or abundance. These data are all numerical, and they are handled entirely
by computer which is used to prepare the information for each species
for map making.

However, we also have about 300, 000 individual record cards, which
are used to record information about a species at a locality in greater
detail and are used particularly for rare or critical taxa. Current cards
are 80-column and could be used as computer input, but the majority,
written on 40-column cards in the period 1956-1968, during the preparation
of the 'Atlas of the British Flora' and the 'Critical Supplement' cannot now
be used as computer input. These records have two main functions: to
supply data for map making and to act as a store of information about the
species. Our experience in running a vascular plant data bank for nearly
20 years is that 95% of the questions we are asked are either species or
locality oriented: 19 questions out of 20 can be answered as long as we
have our data organised in 2 files - by species and counties within each
species, and by grid square. If this is the case would we be justified in
setting up an elaborate and costly computerised system to answer the

other 5% ? Our answer is 'No'. We feel that as long as the data are available by species or locality and a copy of the complete record is supplied, the human computer can then come into use to scan the records for answers to questions about dates or status or habitats, etc. To be honest, many of our records do not include much additional information, and this must be even truer of the majority of specimens in European herbaria. However, we still face the problem of converting an input of two kinds of cards, 40-column and 80-column, into a two-way index.

The simple solution would be to punch the cards we receive and make duplicates. However, there is no 40-column card system which will do this. It must be remembered that both the 40- and 80-column cards sent in by our recorders are primary documents - they may carry much handwritten data. Some of this can be lost during punching, and if errors are made the whole card will have to be transcribed - usually by an unskilled operator who does not understand all the terms used, guesses at the words written and produces a nonsense. To avoid such losses these primary documents are now only punched for species and county number: this takes up only seven columns of numerical punching so that errors are minimal and, on our new 80-column cards, these fields are at extreme ends of the card outside the area designed to be written in.

So, our primary index, by species and by counties within species, is largely unpunched - and the data are not machine readable unless we punch from these a second set of cards, all 80-column. If we had these we could easily prepare duplicates to be sorted into grid reference order and by species within each 10 km square. But the question has to be faced - is it worthwhile punching 300,000 records from cards in a variety of handwriting, all of which will have to be edited by a scientific officer before being handed to a botanically unskilled operator, so that in the end we shall be able to produce tabulated lists which will inevitably be a simplification of the complete record, reduced to fit the 80 columns available? And in addition we shall have 3 cards for every record with the storage problems that implies. The answer we have now reached is a qualified 'No'. We now believe that the best solution is to use microfilm and microfiche.

For several years we have been deeply concerned that the original material we hold is unique: if a Phantom Bomber from nearby Alconbury should crash on the site, or an electrical fault in our wooden buildings reduce the Biological Records Centre to ashes the accumulated data of thousands of British biologists over 20 years would be lost for ever. We

feel that every record which comes into our office should be
photographed into microfilm and the film stored elsewhere (The British
Museum (Natural History) in London has agreed to take the material).
At the same time we find the demand continuously growing for copies of
all data which we hold for counties or similar areas such as national
parks, or local authorities. We have encouraged the setting up of
county biological records centres in Britain and over 20 of these now
exist.

If we need to copy our data for safety reasons, and there is a
growing demand for copies, we only have to find a way of sorting the
film into species and squares (which we can use to define any area) to
cover all our requirements.

Needless to say this can be done. The card index of original records
sorted by species and by counties within species will be filmed in that
order. Three copies of the film will be prepared. One to be sent to the
British Museum (Natural History). The second, already largely in order,
will be fed by machinery, a 'reader/jacket filler', into microfiche, each of
which will hold the data for about 60 cards. Labelled fiche will be stored
in species order, and all enquiries about the data we hold for species will
be sent out by making a duplicate on a 'fiche duplicator'. If there are a
large number of records covering several fiches and the enquiry is
restricted to a particular county only the fiche covering that county
need be sent. But, as the cost of copying one fiche (5p) is the same
whether it is full or empty, it is cheaper to pack the fiche and supply
more data than the enquirer needs. The third film is the tricky one.
The operator will have to scan each frame, on a screen which enlarges the
piece of film to the size of the original, read the grid reference of the
record, select a jacket and insert the piece of film. It will mean manual
sorting of 300, 000 records, but before you gasp in incredulity I ask you
to remember what we are saving.

There has been no editing, no punching, no verifying, no rewriting of
errors, no tabulating of data. The cost of buying and storing 600, 000
cards will be saved as well as large amounts of postage.

The total cost of the equipment to undertake this work is £4000, which
includes a system for producing hard copies at the original size from any
frame on the fiche – and requests for particular information from
questioners not possessing a 'reader' themselves could be answered in
this way. However, most users would be expected to acquire a 'fiche
reader' for £100 and they can, if they wish, purchase their own

reader-printer for less than £500 to make their own hard copies.
There is the problem of additional records but, as we shall not commit the records for a group to film until the editing and mapping stage, the number of additions compared with the original will be small and the scanning of the end of a species or square list for these will not be a great burden for the user.

Data Processing In European Herbaria

Has the system described above any relevance in solving the problems of data processing in the herbaria of Europe?

First I am sure a case could be made, particularly for type material, for a photographic record being made, and copies of the film being stored elsewhere than in the herbarium. In this way further tragic losses, such as those which took place during the last war, can be partially overcome.

Secondly I was constantly surprised when listening to earlier lectures at this meeting with the way in which the specimen was separated from the label. I have always assumed that the best answer an herbarium worker could receive to his questions would be the instant delivery of all the specimens he required, beautifully arranged in some geographical order within the folder: if he could also inspect a complete collection of plants from any area of the world, or find out what an herbarium contains for that area he would be overwhelmed with delight. Is he going to be equally delighted with a print-out of those labels, after they have been transcribed by someone who may have misinterpreted the handwriting, and can he be sure that the original identification was correct (not, as we have heard, in 25% of the cases apparently if your source is the Edinburgh herbarium)?

The way the specimens are already arranged here at Kew is adequate as long as you are working at Kew and the same may well apply to other European herbaria - what we really want to do when we are working in an herbarium is to have access to the specimens in Paris or Vienna without having to go there, or at least to look at the specimens before deciding which we would like to borrow. This could be done, on film, if you would agree that the immediate problem is to provide lists by taxonomic group and by geographical areas. It will be necessary in advance to agree on only two things, a standard list of plant genera and subdivisions of geographical areas. And we are very fortunate because both these now exist. Index Nominum Genericorum is virtually complete, whilst Geocodes and maps, prepared by Sydney Gould, are available for the whole world at the country and province level, and even cover the oceans.

I suggest that one way of beginning some flow of data between the larger herbaria with their millions of specimens would be for the collections to be photographed by genera and by species in alphabetical order, as labelled, within each genus. It would only be necessary to ensure that every specimen was clearly marked with an herbarium identity device and the country or province code before filming took place. I would certainly not try to add grid reference or similar device at this stage. I would leave this to the research worker who would do it once only when he had all the material assembled: often (in Britain anyway) many gatherings are from the same place.

Three copies of the films would be made. One, uncut, could be put into an archive store in another place, a second would be cut up and pocketed in fiche for each genus, whilst the third would be pocketed in fiche for each country and province.

All this work would be done by technicians. Very little mental input would be involved at this stage. I think it important to remember that only a small proportion of our herbarium material is used by outsiders. 30,000 specimens are sent on loan from Kew in a year - less than 1% of their total holding. Can we afford to invest mental effort into specimens which may not be used in the next 100 years? Problems of identifying place names on specimens from Veracruz are clearly more easily answered by, and may indeed only be of interest to, those writing a Flora of that province.

One big advantage of this system would be that any European herbarium which could afford £4000 for the equipment could start producing and distributing copies of their material within a few weeks without employing any specialist staff. It might be possible to organise a scheme for dealing with the major herbaria following a definite taxonomic order over a period of years in much the same way as the Flora Europaea Mapping Project plans to complete its work (though I hope the filming could be done in less than the 65 years which that may take).

If to film all the material is too grandiose a scheme, a modest but invaluable start could be made by agreeing to film all the type material held in each herbarium. Filming could be done on a contract basis so the expense of the major item, the camera, costing ₤1560, could be avoided. All the film could be sent to one centre such as Kew which could prepare and duplicate fiches for each genus.

The advantage of starting with types would be that besides the obvious value of the product to all taxonomists it would act as a pilot study to the more extensive use of microfilm and microfiche in European museums.

Conclusion

To sum up I would like to make three points:

1.   In my experience computerised information retrieval systems are fine as long as you are not working to a time-table, do not have to pay for the computer time of development, and have a relatively small number of records - tens of thousands rather than millions - with a high utilisation (the Mexican situation seems ideal in these respects). You may think that when you have the development charges paid for and the bank set up, that you can count costs at so much per record: but machines become obsolete and computers change. In the Biological Records Centre in 1968 we were forced to begin converting all our data from 40-column cards to computer store. All our calculations suggested it would take 2 years, but it actually took $4\frac{1}{2}$ years at something approaching twice the cost. And just as that was complete we found ourselves facing the threat that, because the computer system on which we hire time is overloaded, we could be asked to move elsewhere, with all the rewriting of software that that would involve. There is always something which I might term the 'Concorde' element in computer systems. In contrast, during the preparation of the 'Atlas of the British Flora', when we were using D.P. rather than E.D.P., we were able to maintain our production schedule and ended up in 1964 handing back to our sponsors from a £30,000 grant just £130.

2.   Any operation which involves the collaboration of many individuals or organisations needs to produce some tangible results within a reasonable period, otherwise the sources of information begin to dry up; and this must take priority over methodology. If you are developing a complex system of data handling it is essential that the system is working before widespread appeals for data are made. And of course if too many years are spent in development

without productivity the money from the sponsor may dry up.

3.   Third and lastly, before deciding on the system to adopt and
        assuming you do not have unlimited funds and unlimited time,
        it might be prudent to ask whether it is better to give a 50%
        answer (divorced from the specimen in the form of print-outs
        and listings) to 100% of the questions which a computerised
        system would give, or a 90% answer (a microfilm of the
        specimen) to 95% of the questions, which might result from the
        two-way microfiche index I have suggested. I probably
        exaggerate the figures, but the point I wish to make is that
        there are two elements involved, the ability to retrieve, and
        what you retrieve - the more effective system may be the one
        which gives the higher quotient.

REFERENCES

Perring, F.H. & Walters, S.M. (1962). Atlas of the British Flora.
    London and Edinburgh. Thomas Nelson.

Perring, F.H. & Sell, P.D. (1968). Critical Supplement to the Atlas
    of the British Flora. London. Thomas Nelson.

THE DATA BANK OF THE BRITISH ANTARCTIC SURVEY'S BOTANICAL SECTION

D. M. Greene and S. W. Greene

British Antarctic Survey Botanical Section

University of Birmingham, England

Summary

A description will be given of the use of a coded computer file and an in-clear card index to retrieve data from a medium-sized, special-purpose herbarium. The computer program uses a fixed-field format for selected collecting data, state of specimen information and herbaria holding duplicates. It is possible to sort on all fields and therefore the file can be used for different purposes. The card index contains the full information on every specimen on the computer file and provides an instantly available reference source.

The herbarium contains specimens from all plant groups but is largely confined to one geographical area, the Antarctic, with a small proportion of taxonomically important specimens from neighbouring areas. It differs from a normal herbarium in having a high percentage of unidentified material.

The file is used to retrieve records of identified and unidentified material for taxonomic and phytogeographical work and to plan each year's field work. The existence of such a data bank has provided a means of permanently storing habitat and geographical data without collecting specimens in an area where conservation is important.

In the late 1950's the British Antarctic Survey began taxonomic and phytogeographical studies on the terrestrial plants of Antarctic regions

with a view to writing floras for vascular plants, bryophytes and lichens.
Antarctic regions extend over approximately 1/5 of the earth's surface and
consist of a geographically very diverse area. However, the vascular
plant component is very limited, with only some 70 native species and about
the same number of introduced aliens. The numbers of bryophytes are
variously estimated at between 100 and 500 species and the lichens at
between 400 and 700 species - somewhere between 600 and 1,200 species
in all.

At the outset only a very limited amount of material was available,
it was widely scattered in world herbaria and much of it was only partially
identified. Indeed when the work began many curators and keepers were
unaware that they had any Antarctic specimens in their care. Another major
problem was the total lack of synthesis in the cryptogamic literature which
consisted for the most part of type descriptions.

The situation virtually changed overnight owing to the great upsurge
in biological interest in Antarctic regions immediately following the
International Geophysical Year of 1957-58 and resulted in a rapid influx
of material from ecologists and physiologists who clamoured for names as
well as the near suffocation "by wealth of specimens" of the few cryptogamic
specialists competent or interested to work on these collections.

Traditional methods of cataloguing soon collapsed as collection
registers became annotated beyond recognition, specimen lists and
determination lists kept altering at a rapid pace and the maintenance of a
card index cross-referencing system to the information began to require
more time than was available to examine the specimens.

During 1968 attention was turned to electronic data processing
methodology but everyone consulted had a different idea of what should be
done and the best way to do it; costs appeared astronomical against the
funds available and requests for additional staff were likely to fall on
deaf ears. But a system had to be found which was relatively inexpensive
to purchase and operate, and this implied the use of simple computer
programmes with, probably, the loss of some flexibility. One of the
real problems was deciding what were the essential requirements, and
with hindsight it has been realised that this was largely due to the
unmethodical nature of many traditional methods. If one has not used a
computer its discipline is unappreciated and hence one is unprepared for
the changes required in routine and for the benefits this can bring.

By good fortune, about this time, contact was established with a representative of Ultronic Data Systems Limited who had been working closely with a representative of a commercial bureau - Autodata of Birmingham, later bought by Hoskyns Systems Limited - and between the two representatives a system was worked out which satisfied the major requirements of the work and was within the budget. To cut computer costs the data were to be coded numerically with only a few alpha codes being used in specified fields. The format was to be fixed field rather than free field, and the file was only to be updated and questioned once a month although no limitation was to be placed on the volume of the input or the number of questions. An input machine was essential and if this were a paper-tape typewriter it could also be used to produce in clear (i.e. normal language) information in upper and lower case for a card file and to replicate herbarium labels.

The paper-tape typewriter currently in use is an Ultronic Data Systems 5000 Series machine with 2 readers and 2 punches and a data select system operating on 13 channels. It is coded in ISO to suit the International Computers 1900 Series computer used by the bureau. The computer program is written in COBOL and consists of a file maintenance routine and a retrieval system. The retrieval system is an ICL Find 2 package. A detailed account of the data bank and the retrieval system has been provided by Greene (1972).

The system starts with the production of the card file. As each specimen is received in the herbarium it is catalogued by typing a 15 x 10 cm. index card on the Ultronic using a machine programme tape which governs format, data selection and insertion of information; at the same time 2 paper tapes are generated. One tape is stored separately and is for emergency use should the card file be destroyed or damaged, and the other is used for the replication of all subsequent cards and herbarium labels. The cards and paper tapes are stored in collection order or, in the case of specimens borrowed from other institutions, under the name of the holding herbarium.

The coded information about each specimen on the computer file is prepared from the cards so that the presence of a card in the card file indicates the presence of a record on the computer file and vice versa. For each record the following fields are used: name of collection, number of specimen, generic name, specific epithet, name of collector, date collected, name of determiner, broad location i.e. geographical region or island group of origin, definite location i.e. a metric grid reference or latitude-longitude co-ordinates, whether the record has been published

or not, the reproductive state of the specimen, and the names of the
herbaria to which duplicates have been sent. Individual records are
identified by means of a file number which is a combination of the
collection name and the serial number of the specimen, the file
maintenance programme rejecting any duplicate entries.

The coded information to update the computer file is punched on to
paper tape and is arranged in one of three ways:

(i)   New records, i.e. all the collecting information going on file for
       the first time, together with a determination, which may be
       a final name, as in the case of a specimen from a world
       herbarium, or a preliminary name if the specimen has only
       recently been collected and awaits detailed study.

(ii)   Updates, i.e. additional information being added for a record
       which was not available when it was first input, e.g. a
       specific epithet where the preliminary determination had only
       been as far as genus, the name of the determiner or the
       reproductive state of the specimen - particularly important
       for cryptogams - and the names of hernaria to which
       duplicates have been sent. An entry is also made to indicate
       if the record has been published and in this case the card will
       bear the full reference.

(iii) Deletes, i.e. the collection name plus the serial number which
       results in the record on file being erased.

A report in 2 parts is received after each monthly update. This lists
all the records received and indicates rejections due to format errors
while the input validation lists duplicate records, mismatched updates
and deletions, i.e. any entries which are not transferred to the master
file. The file is questioned immediately after updating and it is possible
to sort the data in all fields and to have records arranged in any sequence
within fields.

Using this relatively simple system the time taken for herbarium
chores has been considerably decreased and there has been a substantial
improvement in accuracy. For example all specimens with the same
collecting information can be treated together. In this case only the first
card is typed and the tape produced is used to generate the remaining
cards with only the number and the determination being inserted manually.

Therefore only the first card needs checking fully. Herbarium labels are produced in large batches when a genus has been revised, the process being almost totally automatic with the original tape providing the collecting information and a tape in reader 2 providing the determination, determiner and the date the specimen was determined. Again checking is reduced to a minimum.

The computer file has proved invaluable as an aid during taxonomic revisions in that all records preliminarily determined to a genus can be recalled either arranged by geographical area or collector or any other desired combination. Determined material can also be arranged in any required sequence, for example geographically for the preparation of distribution maps. So far no attempt has been made to use computer mapping since geographical survey of the area is still inadequate and thus maps are frequently being altered and, moreover, the amount of time spent plotting by hand is minimal.

The file has also proved very satisfactory for permanent storage of field distribution information. For example for South Georgia, where the vascular flora presents no serious taxonomic problems, the distribution and habitat data of species are being recorded from information in field notebooks of reliable observers without having to administer a large number of specimens which are not needed for any other reason; this method is, therefore, both providing useful information on species performance and helping to conserve the flora.

When planning field programmes to extend the botanical survey, the computer reports have proved invaluable to pinpoint areas where either too little or no work has been undertaken, and as the input of data increases they will also obviously be of help in the selection of the most suitable areas for working on specific problems.

A number of conventions and features have been introduced to accommodate situations such as material which is mixed or has been re-determined one or more times, type material, etc. As the programme requires a unique file number for each record and rejects duplicate entries, 2 alpha suffixes have been included after the serial number, one for mixtures and the other for re-determinations. Thus an A in the re-determination column indicates the first renaming, a B the second and so on, with the exception of the letter T which has been reserved for type material. Using the suffix T it is possible to construct a synonymy list by questioning the computer twice; in the first instance sorting for all records of types which are present for the given genus and then

selecting for those that have been re-determined by one determiner to
a given species.

The requirement of a unique file number presents difficulties where
early collectors did not allocate a unique number to each specimen and
where duplicates are present in several herbaria. This has been dealt
with by assigning a collection code to each world herbarium and
numbering the specimens sequentially as they are entered. This is not
a totally satisfactory solution as there is no direct cross-referencing
to duplicates although these records can still be sorted by collector,
determiner, locality, etc., and it is usually comparatively easy to
reassemble the duplicates by using the computer report to find the relevant
cards.

Another convention concerns the treatment of geographical data. Two
fields have been allocated for this information, i.e. the broad location
which usually corresponds to political boundaries, and the definite location
which, as mentioned earlier, is a precise statement in terms of numerical
co-ordinates either of latitude/longitude or of a metric grid. For
convenience the figure 9 is entered in the appropriate field where the
collecting data are so inadequate that exact co-ordinates can never be
allocated, e.g. for a specimen simply labelled South Georgia, the broad
location is adequate and a normal code is entered but 9's are inserted in
the definite location field as there is no precise statement of where the
material was obtained. Similarly the figure 8 has been employed where a
location has been given but it is in need of checking, e.g. where the
relevant gazetteers are not available or where it is an old place name and
it would require too much work to trace at the time of input. This use of
the figure 9 for information that is totally lost and 8 for information which
further research may reveal is used, in addition to the location fields, in
the collector, determiner and date fields and is a step towards
Professor J.G. Hawkes' idea of a "credibility or confidence index" for
the records on file.

The problem of how to treat minor collectors has also had to be faced,
there being two aspects to the problem. In the case of odd specimens
given to a botanist, who has almost certainly extracted the relevant
habitat and locality information, the specimens are included in his
numerical series. For example a specimen collected by A. Moss and
given to S. Greene is treated as Greene 275, or whatever the next number
is, but the collector's name is still inserted on the file, i.e. the
specimen is identified by the file number which is the collection name
plus number, not necessarily the name of the collector plus number.

However if desired, it can still be retrieved by sorting on a combination
of the collector's name, locality, date collected, etc. In the second
case where a collector has only obtained 30-40 specimens, or less, and
is unlikely to collect again in Antarctic regions, the specimens are put
into a British Antarctic Survey Miscellaneous Collection. In other words
the material is treated as part of an existing series and a new collection
code is not allocated, but once more, as the collector's name is also
input, the specimens can still be retrieved in this way. All voucher
specimens for cytological data are also incorporated into the British
Antarctic Survey Miscellaneous Series and while the limitation of the
computer programme means that these records are not specially
tagged on the file, the cards indicate the nature of the specimens.

Reference has been made by Mr. R. Ross to associated data about
specimens of which a taxonomist needs to be aware and which, in the
case of the British Antarctic Survey herbarium, take the form of
microscopic preparations, photographs of type or illegible labels, maps
of collecting localities, etc. Since the card file is maintained as the
most up-to-date source of information about the specimens, annotations
are added as appropriate to indicate the nature and siting of the
associated data, e.g. the existence of a slide preparation. In other
words sorting is by the specimen itself (which could easily be a slide
preparation, a bottled specimen, or any other sort of material), the
card being used to indicate the position and nature of any associated
material.

Another problem which has been raised in one of the discussions
concerns the handling of data supplementary to the original label, e.g.
the case where the handwriting alone reveals who determined the
specimen. In this type of situation the card is annotated with an entry
in square brackets but the information is inserted without qualification on
the computer file. This is no disadvantage as the card index is the
prime data source, the computer file being really a sorting mechanism
to direct the researcher to the cards or specimens required. Obviously
a card file of this sort may not be practicable for a very large herbarium but
it represents a convenient place where extra data, cross references, etc.
can be added, thus allowing continuous improvement to the data about a
specimen without posing a computer handling problem. Indeed if access to
the computer file is limited, the annotations on the card are essential for
day-to-day convenience.

Such a combination of card file and computer file has been in
operation in the British Antarctic Survey herbarium for 4 years. The

coding and supervision is undertaken part-time by one of the authors and requires 5-6 days/month. Botanists, not typists or technicians, prepare the data about specimens for input onto cards. The Ultronic input machine is operated by a competent typist working a half day daily for 40 weeks of the year, and during the 4 years some 30,000 records have been entered onto the computer file, many of which have been updated subsequent to the original entry. The cards to cover these records have been typed and filed and approximately 10,000 herbarium labels have been generated. In addition a variety of print-outs have been obtained and the machine operator has prepared for publication lists of specimens examined using a computer report which she decodes directly as she types. In fact the majority of her time is devoted to the support of normal curatorial activities, the input only taking 3-4 half days/month to type.

It would be idle to pretend that there have not been difficulties in getting the system operational. The first year was spent largely in ironing out problems with the computer programme, learning the basic use of the input machine and refining its operation to make it as efficient as possible, and laying the basis for the code book. There was also some difficulty in persuading reluctant taxonomists that a computerised system was better than the traditional lists and counter lists which can be so easily mislaid. It was found to be particularly difficult to overcome the reluctance of taxonomists when there was very little information on file but as soon as the records for one geographical entity were completed, and the ease and speed of working with material for this area could be compared with the results of using the traditional methods still required for areas not yet on file, most were rapidly converted.

There are, of course, disadvantages in the system used in the British Antarctic Survey herbarium. It does take time to code the information, to maintain an up-to-date code book and to decode the reports. Within the confines of one institution and with the present size of file this has not proved serious, but it would be a simple matter, given money, to get the computer to code automatically from in-clear information and to decode it again when producing print-outs so that the user has all the information in-clear. It is equally obvious that a small increase in operating costs would give weekly rather than monthly access to the file - at the moment monthly is adequate but if the enquiry rate were to increase it would not remain so.

REFERENCE

Greene, D. M. (1972). A taxonomic data bank and retrieval system
     for a small herbarium. <u>Taxon,</u> 21: 621-629

DISCUSSION, MORNING, THURSDAY, 4 OCTOBER

This was a general discussion on papers read during this morning.

Dr. R.J. Pankhurst. The danger in putting too much reliance on the human computer is that the data-expertise is lost with the death of the individual.

Mr. J.F.M. Cannon. Is it possible to update the microfiche record discussed by Dr. Perring when new material accumulates or new type-specimens are selected?

Dr. F.H. Perring. Such corrections can be done in our own system and periodic revisions of other systems could be undertaken.

Mr. J.F.M. Cannon. Is the value of the complete revision worthwhile?

Dr. F.H. Perring. This is a matter of detail for each user to decide.

Mr. J.P.M. Brenan. The photographs stored in the two files are in geographic and systematic order. How are these reorganised when additions are made?

Dr. F.H. Perring. Each fiche would be devoted to a genus or section, or to a country or subdivision. Additions would be cut up and added to the appropriate fiche. This is easy if the labelling of the specimens is carefully done.

Mr. L. Ryvarden. There are great advantages in this microfiche system described by Dr. Perring. With special coding on the labels might it not be possible for electronic sorting of these to be done as is done by the Post Office using postal codes?

Dr. F.H. Perring. This is certainly a possibility.

Mr. R. Ross. Index Nominum Genericorum provides a list of generic names but does not distinguish between accepted names and synonyms and is not, therefore, adequate for the purpose suggested although it does provide a good base. Technicians are quite capable of extracting geographical data from labels with the exception of some early 19th century handwritten labels which need more expert interpretation.

Dr. F.H. Perring. This should not prove a major problem as we found that technicians soon become very skilled at this sort of work. Botanists must agree on a list of names to use for retrieval purposes, as they do in herbaria.

Dr. R.K. Brummitt. In recording the geographical distribution of a species a generic thesaurus is unnecessary, but it is necessary to have an acceptable geographic list of country names. This meeting could ensure that this was produced. I don't know if there is an easy method of relating the label data to the geocode. The production of a type register by photographic means - microfiche - can be undertaken in a finite time as was done with the Wallich Herbarium at Kew. The photographic types from all herbaria could be centrally pooled. It would only be necessary to indicate what the specimen is a type of and not what accepted taxon is represented.

Dr. F.H. Perring. These details can be worked out by discussion of the interested parties. The geocode includes a book of maps and thus can be associated with label data. Such a hierarchical code has advantages. A central agency could collect all the photographs and arrange them in some agreed taxonomic order.

Mr. J. Raynal. The herbarium itself, like Dr. Perring's system, is an efficient data file not requiring E.D.P. technique. The specimens, arranged in a combined taxonomic and geographical system, form a two-dimensional matrix which is easy to consult. The idea of a central collection of photographic types is a good one but it will be a great task to define the types.

Professor J.G. Hawkes. The photographic system is admittedly useful but we should not forget that E.D.P. can answer questions of great use to taxonomists. For example, what herbaria have collections of a given collector? What herbaria have duplicates of given specimens? What families, genera and species have been collected from a specific geographic area, country or locality? What habitat references can be obtained from a specific locality from all the specimens collected there? Other questions which E.D.P. could help to solve quickly are what specimens were collected before and after a specific collection so as to supply missing locality data, and which herbaria have isotypes etc. for whole groups. This kind of information is very useful for monographers and for future expeditions.

Dr. F. H. Perring. A separate file could be made for collectors, but I would question the importance of this information in relation to the system which I propose for dealing with types. Such information could often be read directly from the photographic record of the label. An E. D. P. system could easily be set up from the photographic record such as I suggest.

Professor J. G. Hawkes. Such a system would not provide information for other taxonomic groups.

Professor Dr. M. Riedl. While Professor Hawkes's suggestion is desirable, an institution with small staff numbers and limited finances has to be realistic about what it can do.

Dr. D. J. Rogers. We should examine how long a time it takes for the physical handling of the specimens for making the microfiche.

Dr. R. K. Brummitt. The Wallich Herbarium, containing tens of thousands of specimens, was done in approximately 6 weeks.

Dr. D. J. Rogers. It is vital to make an accurate and detailed comparison of the times of different operations together with a comparison of costs. One must remember that even after the microfiche is made, it is necessary to extract the information from this photographic record.

Mr. J. Raynal. There is an advantage in having photographs because it avoids the handling of specimens.

Dr. S. G. Shetler. This photographic record only creates yet another herbarium. The problem we face is indexing the data in our collections.

Dr. G. Panigrahi. The publication of an International Type Register to indicate the names of herbaria where types of vascular plants are located and the collections of type photographs (together with their negatives) at a centrally located herbarium, e. g. Kew, would be very useful to the developing countries. These often have large collections but few authenticated or type specimens and have had to depend on old and out-of-date Floras of their respective regions for identification of their collections. The photographs of types of species they are concerned with, if they are made easily available, either at cost or on basis of exchange of recent collections, would undoubtedly assist in quickening the progress in writing the Floras of such countries.

Dr. J.F. Mello. Could Dr. Cutbill define what he means by standards?

Dr. J.L. Cutbill. A standard is a set of properties which must be possessed by data before it can be claimed that they conform to the standard.

Dr. J.F. Mello. What about standards for colour?

Dr. J.L. Cutbill. The Information Retrieval Group of the Museums Association - I.R.G.M.A. have taken no responsibility for this.

Professor A. Gómez-Pompa. Have you any standards for localities?

Dr. J.L. Cutbill. Political names, longitude and latitude, and grid references are used, but latitude and longitude are recommended.

Professor Dr. M. Riedl. Can you apply your standards to old collections which have only poor data? Is there a means of translating scanty data?

Dr. J.L. Cutbill. This presents no difficulty; a low information content is indicated in the appropriate places. You cannot generate what does not exist.

Mr. J.P.M. Brenan. With interdisciplinary standards might it be necessary to develop this further?

Dr. J.L. Cutbill. At the level at which standards exist there is little difference between the disciplines as they all have descriptive problems in common. Each thesaurus of key words may certainly be different.

Professor C. Kalkman. The standards you describe are common to all international museums. In drawing them up who was responsible, curators of local museums, staff from national institutions, or scientists?

Dr. J.L. Cutbill. The initiative came principally from museum staff dealing with the humanities. It arose because of their awareness of their inability to answer questions from existing data.

Professor C. Kalkman. I find it difficult to see any value in setting up these standards.

Dr. D. J. Rogers. We seem to be getting confused between standards for substantive data and a structure which makes this data into a generally useful system.

Dr. J. L. Cutbill. The system was evolved because a general filing system was required for interdisciplinary museums and existing systems were inadequate.

Dr. D. B. Williams. One problem is that we cannot predict in advance all the data which might be needed.

Professor Dr. K. Walther. There is a system available in which all colours are translated into figures; would not this provide a standard?

Dr. J. L. Cutbill. Yes, if everyone accepts it.

AMERICAN HORTICULTURAL SOCIETY PLANT RECORDS CENTER

R.A. Brown

American Horticultural Society

Plant Records Center, Mount Vernon, Virginia, U.S.A.

Summary

Established in 1970 by a grant from Longwood Foundation Inc. to the American Horticultural Society (A.H.S.), the Plant Records Center (P.R.C.) has made considerable progress towards the development of a national center for horticultural information. The P.R.C. has developed a standardised system for the recording and reporting of information related to plants cultivated within major North American botanical gardens and arboreta. Utilising modern data-processing techniques, the P.R.C. has developed a system of reports or inventories which provide managers of botanical collections with timely inexpensive reports on their living collections. In addition, the P.R.C. offers to its co-operating institutions other services, including keypunching, programming, and statistical processing.

While much of the resources of the P.R.C. is directed towards development of projects related to botanical garden records, the P.R.C. is actively investigating other services. The P.R.C. recently announced its file-researching service to botanists, horticulturists and others. With over 170,000 records within its data files, the P.R.C. can often direct professionals who are seeking information about particular plants to institutions where the desired plant is reported to be growing. The P.R.C. is also actively investigating the field of "word processing", to assist those institutions which are engaged in the preparation of recurrent publications, such as indices, directories or registration lists.

For the A.H.S., the P.R.C. is developing a system to manage membership (subscription) record files. Once established, this system will be offered as a service to other horticultural organizations located at the A.H.S. headquarters.

Introduction

According to the International Directory of Botanical Gardens (1969) there are more than 100 botanic gardens and arboreta within North America. If considered in combination, these institutions are probably cultivating over 1,000,000 plant specimens. It is reasonable to assume that, considering this wealth of plant material, a specimen of almost any cultivated plant might be found in one or more of these collections, if one knew where to look.

Recognising the value of these collections and the wealth of information which they represent, a feasibility study was initiated in 1966 to consider the concept of an International Plant Records Compilation Center, prompted by a report presented at the XVII International Horticultural Congress by Mr Robert D. MacDonald, then Assistant Professor of Forestry and Director of the Arboretum at the University of Tennessee, who became the first Director of the Plant Records Center and served in that capacity until 1972. Through the unrelenting efforts of Dr. Richard A. Howard, The Director of the Arnold Arboretum, funds were received for this study from the American Association of Botanical Gardens and Arboreta (A.A.B.G.A.), including donations from eleven of its member institutions.

While the A.A.B.G.A. was involved in this feasibility study, the Long Range Planning Committee of the American Horticultural Society (A.H.S.) was considering a data processing center as one of its objectives. An effort to co-ordinate the activities of the A.A.B.G.A. and the plans of the A.H.S. was begun on January 28, 1967, by Dr. David G. Leach, then Chairman of the Long Range Planning Committee.

The result of these efforts was the formation of a joint A.A.B.G.A.- A.H.S. Plant Records Center Advisory Council in October, 1967. Meeting at Longwood Gardens on December 3rd of that year, the Council decided that a grant proposal would be submitted to Longwood Foundation Inc. and, if accepted, a Plant Records Center Pilot Project would be established under the administration of the A.H.S.

On July 15, 1968, the grant proposal was indeed accepted, and the Plant Records Center Pilot Project became operational. Longwood Foundation provided $90,000 to the A.H.S. to finance the two-year pilot study which would utilize the accession records of Longwood Gardens as its data base.

The pilot project was so successful that, when again approached for financial assistance in 1970, Longwood Foundation agreed to provide $1,835,600 to expand to the functional level the Plant Records Center (P.R.C.) of the American Horticultural Society, contingent upon acceptable annual reports of progress.

On April 1, 1973, the A.H.S.-P.R.C. concluded its third year of operation - its sixth year of existence - after an investment of over $760,000, and after more than 170,000 plant records had been processed. With the notable changes achieved during the past 12 months, fiscal year IV (April 1, 1973—March 31, 1974) promises to be a significant step towards the ultimate success projected for the ten-year program, hallmarked by the move of the P.R.C. to its new home within the magnificent A.H.S. headquarters in Virginia. Here, with a new staff of assistants and equipment operators, trained within a framework of updated operating policies and principles, the P.R.C. anticipates a growing program of services.

Program Objectives

The P.R.C. has three principal objectives, the first of which focuses on the establishment of a standard system for the recording of information relating to plant accessions. To meet this objective the P.R.C., in its early pilot study stages, reviewed over 100 accession record systems. As a result of this study, a "standard" accession card was developed that would satisfy two design requirements. Firstly, the card was designed to provide for the consistent, uniform reporting of information considered to be of principal importance to most institutions. Secondly, the card was prepared as a two-part form (original and carbon copy), the original file copy to be maintained by the institution preparing the record, the carbon copy to be forwarded to the P.R.C. for processing into the data bank. To assist the recorder, the P.R.C. provides an instruction manual which describes in detail how the card is to be filled in, and how changes or deletions are to be submitted to the P.R.C. for records previously processed. Currently, 37 botanic gardens and arboreta are utilizing this accession card system. Using many of the suggestions from these users, the standard accession card has gone through five minor revisions.

The second and perhaps most obvious objective of the P.R.C. is
to develop a·data bank of the cultivated ornamental plants of North
America. Currently, 33 botanic gardens and arboreta are co-operating
with the P.R.C. in this program. To date, the P.R.C. has totally
processed 18 collections, representing some 170,000 accession records,
and an additional 4 projects are at varying stages of completion
representing an additional 20,000 accession records. The information
recorded and processed on any one accession consists of:

> garden (collection) code
> scientific name
> authority
> accession number
> family name
> common name
> source
> location within the collection
> country of origin
> number received
> how received
> date received

Frequently, additional information will be reported relating to:
number of herbarium vouchers prepared, if any; propagation
information; descriptive information; and collection data.

Once processed, any one or more of these items of information may
form the basis for the preparation of a computer-generated report
(data listing or print out) representing one or more collections.

The third objective of the P.R.C. is, perhaps, its most significant
one. Through use of its modern equipment and data handling techniques,
it is establishing an information center designed to service three current
requirements:-

1.   The P.R.C. offers an accession recording and reporting system
to botanic garden and arboretum management. As a records-management
consulting organization, it offers assistance to meet special record needs.
Recently developed computer programs have proven to be valuable tools
for garden management by providing detailed reports on the living
collections on a timely, economical basis.

2.   The P.R.C. offers special assistance to institutions which require additional services beyond the normal scope of the P.R.C. program. By using available statistical analysis computer programs or other "packaged" computer programs, the P.R.C. can process data files other than plant accession records, and assist in the analysis of special research projects. In this capacity, the P.R.C. offers the benefits of computer processing to any of its co-operators.

3.   The P.R.C. disseminates information to the scientific, professional, and amateur communities, which draw on the botanical disciplines for information and materials. The P.R.C. has recently announced a researching service for scientists, horticulturists, and others; it will scan its data files, which have over 170,000 records, for information relating to desired plants. More often than not, the P.R.C. is capable of directing the researcher to institutions where the desired plant is reported to be growing, and can help to guide the researcher to additional sources of information. Many of the reported institutions can further direct the researcher to commercial sources, if actual distribution of plant materials is not permitted. For such services, the P.R.C. has established schedules of charges, dependent upon the nature of the request.

Recently, the P.R.C. has begun to investigate the field of "word processing". For botanical or horticultural organizations involved with preparation of recurrent publications, such as registration lists or directories, the P.R.C. believes it can be of considerable assistance by preparing the manuscript for computer processing. The principal advantage of being computerised relates to ease of sorting and combining additions or changes to the basic text. Combined with the capabilities of high-speed printing with several optional character fonts available (including upper and lower case characters), the P.R.C. believes that this will reduce much of the cost of setting type, preparing proofs and editing. Additional investigations with a case example will be undertaken in coming months.

Outline Of Operation

Information is received by the P.R.C. in two ways. One way is through the use of the standardized card as filled in by a garden. Cards are sent in only when the accessions have achieved a degree of relative permanence in a collection. These cards are sent in batches, usually according to prearranged schedules. The second way in which the P.R.C.

receives information is by P.R.C. staff visiting a garden and
microfilming records pertaining to the living plant collection. After
microfilming has been accomplished, the garden utilises the standard
P.R.C. accession card for all new accessions.

When information is received through the use of the P.R.C.
standard accession card, the following steps are taken in the preparation
of the information. First, each card is given a sequential identification
number (RECORD ID). This operation is necessary so that punch cards
may be collated by the computer, at a later step in the operation, to form
complete records, regardless of the sequence in which the punch cards
are submitted to the computer. In addition, this number serves as a
convenient reference code so that an original P.R.C. accession card
may be readily located and referenced to any given punch card. Through
the use of a mechanical stamp, operated by hand, about 10,000 records
a day can be numbered by one person.

After numbering, the accession cards are edited by the appropriate
P.R.C. staff to ensure that all data have been entered correctly. At
this point, certain codes are entered on the record to indicate what
various kinds of information (other than the required basic information)
are present on the record. Next, the basic information entered is
keypunched and prepared for processing. After keypunching, the punched
cards are key verified, thus eliminating over 90% of the errors created
during the initial punched card preparation. Once keypunched and key
verified, the punched cards are "fed" to a card reader attached to the
computer terminal (a device, which, through the use of a special
telephone, "communicates" to a computer located in New Jersey). The
card reader senses the punched holes of the cards, transmits the
information to the main terminal hardware where a stored computer
program converts the punch-coded information into instructions that
guide the P.R.C. line-printer (also connected to the terminal) to print
listings displaying the information recorded on the punched cards. The
terminal, with its attached card reader and line-printer, is capable of
reading and printing the information on the punched cards at a rate of
approximately 300 cards per minute. The listings are then given a
cursory edit for any remaining errors.

When a garden's records are to be microfilmed by the P.R.C., a few
additional operations are required. Before microfilming, all appropriate
staff of the garden are interviewed to determine just what records are
present pertaining to the living collection of plants. Required records

are selected, and each record to be microfilmed is given a sequential
identification number (RECORD ID) assigned for the same purposes as
with the standard accession card. Using an automatic camera (which
photographs onto two rolls of film simultaneously) with an automatic
document feeder, some 15,000 cards an hour may be photographed by
one person with an assistant. The camera photographs both sides of
the records at the same time, and permits about 6,000 records
(5 x 8" size) to be recorded on a single 100 ft. roll of film. After
filming, each roll of film is sent for processing. The usual time
required to receive the processed film is three days. Upon receipt
of the processed film, one roll is filed as a safeguard in the event of
loss of either the original records or the duplicate roll of film.

Processing of the information on the microfilm begins by keypunching
the data. As with the case of the standard accession cards, each basic
plant record requires the use of five different punch cards. Only one
type of card is keypunched at a time by a particular keypunch operator.
One operator may do only Card 1's, while another operator is working
on Card 2's for the records of a particular garden. The operators obtain
the necessary information from the garden accession records through the
use of microfilm readers, which are positioned beside each keypunch.

After the punched cards have been prepared and edited, the records are
ready for final processing through the P.R.C. data terminal. The terminal,
once connected to the remotely located computer facility, permits the
information that is read from the punched cards by the card reader to be
transmitted directly to the computer.

For processing the plant record data, the P.R.C. has developed a
system of computer programs designed to perform two basic functions:
file maintenance and report generation. To ensure independence from
any one vendor or computer system, all of the P.R.C. computer programs
have been written in COBOL language and are fully documented. Each of
these functions may, in fact, require several computer programs, but they
have been linked together in such a fashion that the P.R.C. operator need
only concern himself with one or the other function.

File Maintenance Subsystem (see Fig. 1)

The main objective of the file maintenance subsystem is the
production of a garden master record file, suitable for extraction and
reporting of plant record data. Each time file maintenance is

performed a new master file is created, accompanied by a series of
"Transaction Error Reports", which list errors detected by the edit
program from the input data.

The file maintenance subsystem consists of three programs. The
first sorts the input transactions (new record card-sets, deletion or
change cards) into record identifier (RECORD ID) sequence, then
edits each transaction. This program eliminates the need to collate
the punched cards prior to processing, and reduces pre-processing
editing. The second sorts the valid transactions, i.e. those which were
not rejected by the editing program, into accession number sequence.
The third is responsible for merging the valid transactions with older
records acquired from a previously processed master record file, and
for creating a new master file. When merging new records with older
ones, this program edits for potential duplicate records. Any new
record bearing the same accession number as a previously processed
record will be rejected and listed as a duplicate record in the "Update
Error Report". In addition to displaying the new record, the error report
shows the older record as well, enabling the P.R.C. Editor to determine
what corrective action is required.

The "Transaction Error Report", as prepared by the edit program, has
been designed to display an error message, the action taken by the
program, plus the full 80 character card (transaction) image. Program
action will take the form of either rejecting the field in error (setting the
field to blanks), dropping the entire card(s) from further processing
(as in the case of a missing accession number), or no action, i.e., the
field has been noted in error but has been permitted to remain with the
record. Since left-justification of all fields has been programmed into
the system, invalidly justified fields are not reported as errors.

The "Update Error Report" lists all duplicate records, plus all
change or deletion transactions where no corresponding master file
accession record was found. An additional update report lists all
changes by displaying the field name, previous contents of the field,
plus accession number. For deletions, the entire record deleted is
displayed.

At the conclusion of a file maintenance run, four totals are displayed,
indicating the number of records read from the "old" master file, the
number of "new" records added to the file, the number of records deleted
from the file, and the new, total number of records written to the "new"
master file.

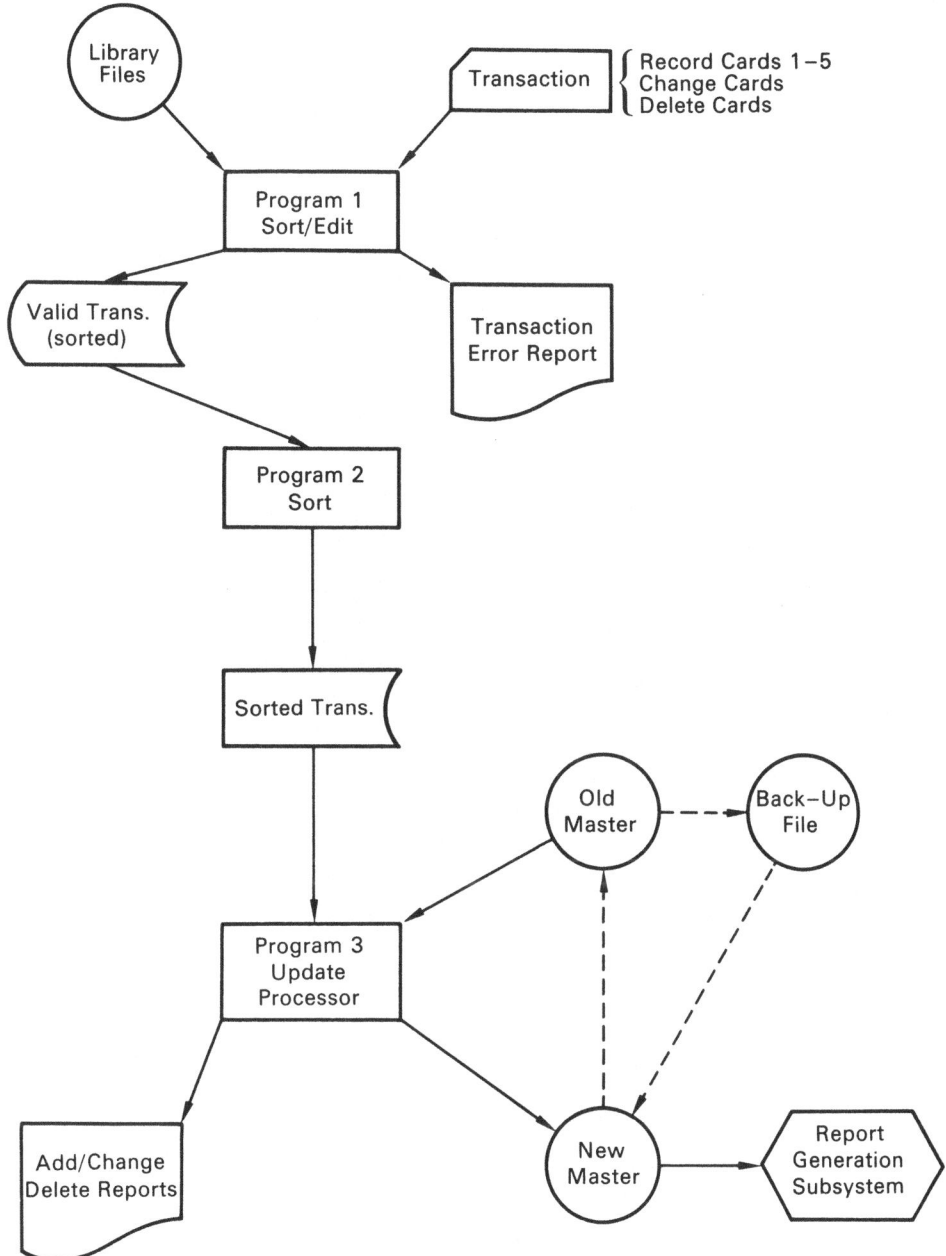

Figure 1. Plant Record Processing System: File Maintenance Subsystem

Report Generation Subsystem (see Fig. 2)

At present the P.R.C. has six computer programs which generate reports, inventories or data listings. Five of these programs were designed to produce specific listings which were developed primarily for use by participating gardens as an internal report system. Whilst the information contained in these listings may be of use to other gardens, they are intended primarily to be used as management and research aids by the staff of the garden whose records are reported. The first of these programs produces a report displaying all the family names (in alphabetical order) for the records contained within the file. Within each family name, plant names are displayed in alphabetical order, with a number indicating the total number of accessions within the file bearing that name. At the end of the listing, the total number of taxa within the file is reported, plus the total number of accessions.

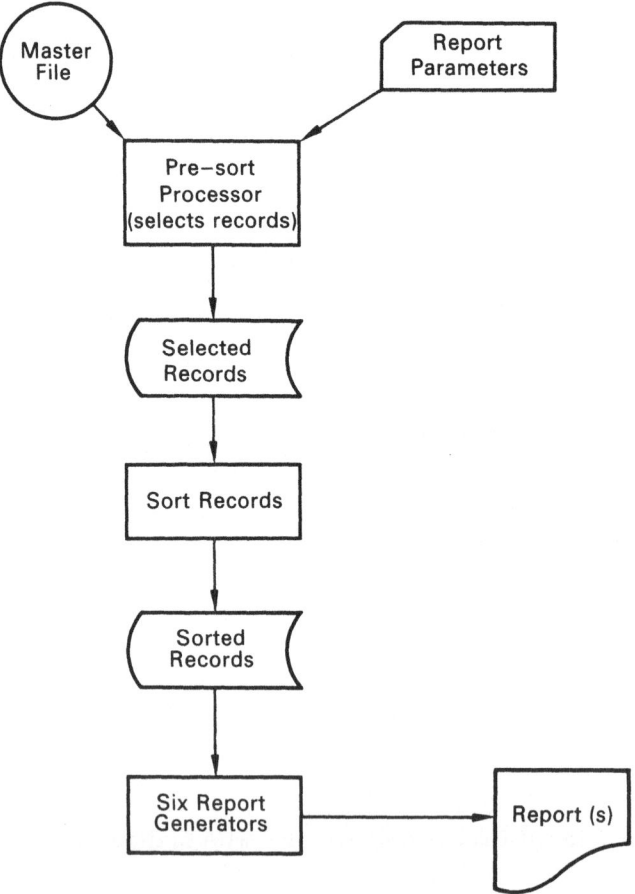

Figure 2. Plant Record Processing System: Report Generation Subsystem

The second program prepares a report much like the first, except that family names are not reported, and displays all of the totals indicated for the first program, but presents the taxa in alphabetical order.

The third program reports every accession contained within a file. This program lists all the records in numerical order by accession number, and displays plant name and family name for each record. In addition, at the end of the report, totals are presented which display the number of accessions acquired per year, in chronological order.

The fourth program prepares an inventory by location data. For each location designation, plant name and accession number are listed for plants reported to be growing within that location. Within each location, plants are listed in alphabetical order by name. In addition, totals are reported showing the number of taxa and number of accessions growing within each location area.

The fifth program is designed to produce, on specially prepared paper stock, new accession record cards. Where a second file of records is desired, or when a new file is needed to replace an out-dated file, this program prepares, in alphabetical order by plant name, new 5 x 8" record cards which are compatible with the P.R.C. standard accession cards.

The sixth program was developed to handle special reporting needs. This program, which is modified at the time of use, permits an infinite variety of listings to be produced. Although this program, like the others, mainly services the information requirements of co-operating botanic gardens, it is also used to service specific requests for information received from institutions other than botanic gardens and arboreta, and from individual research workers.

Cost Of Services

The P.R.C. has established specific schedules of costs for user services, based upon the nature of the services provided. Under the provisions of the Longwood Foundation Grant to the A.H.S., there is no charge for documentation of the living accession records of co-operating, non-profit, educational, charitable, or governmental botanic gardens and arboreta. This operation includes microfilming all pertinent accession records (when feasible to do so), and preparing them for data processing.

The same services are offered to individuals, gardens, or other institutions which utilise the services of the P.R.C. on a continuing basis but which do not qualify as being governmental, non-profit, charitable, or educational. For these, the P.R.C. charges a fee of from $80 to $100 per 100 records involved.

Co-operation with the P.R.C. requires that the user purchase P.R.C. standard accession cards. The cost of the cards is $5 per 100 cards, and includes, for new co-operators, the cost for one copy of the "Accession Manual". Once a garden's records have been microfilmed, the P.R.C. accession cards serve to supply the P.R.C. with information on subsequent accessions. The "Accession Manual", which may be purchased separately at a cost of $20, outlines the use of the accession cards. Revisions of or supplements to this manual are provided without charge as the P.R.C. system develops.

Conclusion

While the basic program of acquiring new co-operating botanical gardens and processing of their records is continuing, additional services are being investigated with the aim of broadening the overall capabilities of the center. One such service is a membership (subscription) file processing system. Once proven, this system will be offered as an available service to horticultural organisations located at the A.H.S. headquarters. For its botanic garden co-operators, the P.R.C. offers many data processing services including keypunching and special computer programming. Though these services can be acquired from other agencies, the P.R.C. believes it can provide them at cost-competitive prices with an added bonus of experience in handling botanical data. As an in-house aid for P.R.C. staff, with potential value to certain co-operators, the P.R.C. is also preparing a directory to the U.S.D.A. Plant Introduction records. This directory, when completed, will guide the user to specific U.S.D.A. plant introduction numbers for recorded taxa.

While these activities may appear to be somewhat removed from the original task of the center - that of being a repository for botanic garden records - it is believed that they are compatible with the overall concept of the center: that the P.R.C. should become a national center for horticultural information.

REFERENCE

Fletcher, H.R., Henderson, D.M. & Prentice, H.T. (1969)
International Directory of Botanical Gardens, II. Regnum Veg. 63.
II Int. Bur. Plant Tax. Nom, Utrecht.

DATA STANDARDS FOR COLLECTION-HOLDING ORGANISATIONS

J. L. Cutbill

Sedgwick Museum, Cambridge, England

Extended abstract of paper read at the meeting

The Information Retrieval Group of the Museums Association of Great Britain (I. R. G. M. A) was established by the Association to promote the development and use of automatic data processing for collection documentation. The group recognises a need for a system capable of integrating diverse data which would facilitate the creation of **interdisciplinary data bases**, the exchange of information between organisations, and the central development of computer program **packages**. The group has therefore been working on a single standard for collection documentation. The Council of the Association has declared its intention of promoting this standard for use by British collection-holding organisations, including herbaria. The International Committee on Museums working party on museum documentation has decided to adopt the British approach as the starting point for developing an international standard.

The usual way to develop a standard is through a listing of all the different kinds of data that occur. This approach is unsatisfactory even in a single scientific discipline such as botany. It leads to over-definition in the standard, difficulties of overlap between categories of information and a rapid escalation in number of categories. In a wider context it fails totally as I. R. G. M. A. and others have discovered. The procedure now adopted by I. R. G. M. A. starts from the general and works towards the particular. A number of data categories such as name, identification, time, place, person, and description are defined as a set of rules produced for combining these elements into statements about objects and concepts. The result is a data format that in no way restricts what may be said, but the data is rigorously organised and thus machine-processable.

The work now in progress involves exploring the information requirements of different disciplines and working out ways of expressing their requirements through the general format. Thus standardisation proceeds from the top down and detail is left unstandardised. This is a natural way of proceeding in that it is not the job of managers of collections to dictate standards not already agreed by a consensus of experts in a particular discipline.

In parallel with this work, programs to be used on data organised to the standard are being developed at Cambridge. These programs are based on the CGDS package. This is a Data Base Management System designed explicitly for use on data related to collections. It is already in use for various projects in the Department of Geology at Cambridge and for the computer-based collection documentation activities at the British Museum (Natural History). Recently it has been adopted by the Scientific Periodicals Library at Cambridge to replace a package previously used for construction of union catalogues of library holdings.

INFORMATION MANAGEMENT AND USE OF TAXIR IN HERBARIA*

D. J. Rogers

Department of EPO Biology

University of Colorado, U.S.A.

Summary

TAXIR (TAXonomic Information Retrieval) is the acronym for a computer-oriented information retrieval system designed to be used in taxonomic work of all types. The system is modelled on efficient working procedures already established in the discipline of taxonomy, employing concepts familiar to most taxonomists. It is a generalized system, useful for any type of data or description of either specimens or taxa. The TAXIR system may be used by the herbarium for efficient record-keeping on loans and accessions, etc., and may also be used by individual taxonomists in the herbaria for the purpose of recording all types of classificatory information used in taxonomic work. It is thus a general-purpose system.

TAXIR programs have been widely distributed in the U.S.A., Europe and U.K. The program varies from place to place, and each is not equally well-developed. There are restricted versions (less general applicability) and more generalized versions, so that TAXIR has become a generic acronym with several "species".

As with any new device or method, it is necessary to have instruction in the proper use of TAXIR (or any other computer system). Adoption of the programs should be preceeded by a short period of training to familiarize users of the system with its capabilities. Also, it is necessary to understand that TAXIR makes communications within and

*Paper No. 34, Taximetrics Laboratory.

between herbaria more rapid and efficient, and the proper application to accomplish the goal of efficient communication implies <u>systematic</u> applications. Instruction aids in accomplishing this goal.

Adoption of TAXIR by any one institution does not prevent communication with other systems adopted by other institutions. Conversion of information between one computer program and another is easily accomplished. Those institutions already working with another system (The Smithsonian, Cambridge, etc.) may easily exchange information by simple computer conversion programs.

The purposes of this presentation are twofold: 1, to discuss the problems of management of information in herbaria and 2, to describe TAXIR (<u>Tax</u>onomic <u>I</u>nformation <u>R</u>etrieval) as a software package which may be used for the purposes of information management. E.D.P. is a term which is so all-inclusive that it is necessary to restrict the definition for our purposes. The objectives of this meeting (as I understand them) are firstly to discover whether or not electronic data processing methods should be employed for storage and retrieval, and secondly to discover if there be any standards applied to (mostly) label information from herbarium specimens in various herbaria in Europe.

1.   Information Management

It is useful to imbed the concepts of E.D.P. in a framework, to know precisely the level of discussion at any particular time. The diagram sketch and the concept presented below is by no means new, but one that is fairly common. We have labelled the concept as the "knowledge triangle". (See Fig. 1)

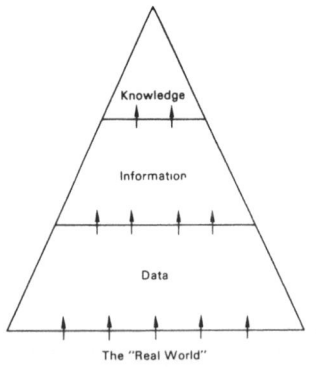

Figure 1. The knowledge triangle

The triangle has three essential sections. The base area of the triangle contains data, and that part of E.D.P. applicable to proper handling of these data is generally referred to as information storage and retrieval. In this portion of the triangle are found all of the measurements or descriptions of the "real world" (in this case, specifically the collections in herbaria). Data may describe anything about the organism or related to the organism represented by the specimen or collection in the herbarium. "Label data" frequently contain types of information that identify the individual specimen. We are all aware that the herbarium, or label, name is not necessarily the most accurate one for the organism. However, it does represent some means for identification of that individual specimen. There are other data on the herbarium label which have the same function, e.g., the name of the collector and the collector's number further specify which particular specimen we refer to. In addition to these types of information on the label, there are frequently the date of collection, the geographic locality of the specimen, and, on more recent materials particularly, various habitat descriptions. We may also find notes about the use of the plants. Although the data on the specimens may indeed be erroneous in one part or another, we still use the data from the specimen as a means to locate and specify that particular specimen. We will also, in the normal work of a taxonomist, use these data as we process information to produce floras or monographs, and we frequently modify some of the information contained on the label, but we have adopted rather well-defined rules for making such modifications. The initial data on a label is never changed - rather we annotate the specimen using separate labels, and many of you are familiar with specimens that have more annotations than there is plant material.

In the knowledge triangle, the next section of the triangle is labelled "information". In other words, we move through the triangle from data to information - correlating the data by various techniques and, in the process of correlation, using other E.D.P. systems to aid in these types of correlation. Many statistical analyses fall into the category of data correlation and the results of such analyses provide the "information" part of the triangle. Thirdly, knowledge is at the top of the triangle. By adroit use of both data and information, we eventually produce knowledge and, in the particular case of the taxonomist, knowledge resides in the taxonomic reports, whether a note, a monograph, or some floristic study. In the production of knowledge from data and information, we employ another set of E.D.P. methods, frequently referred to in the computing milieu as "clustering techniques". In these techniques (of which there are several) the various pieces of information (referred to generally as characters) form the basis for clustering the specimens into taxa, and from the technique we derive conclusions about the organisms, the classification thereof, and the hierarchy of taxa contained in our study.

Each step in the study of a specimen requires a special technique for use of the data or the information, and each of these steps requires some management technique. Not only in the actual scientific study of the specimens do we concern ourselves with management, but we may also consider that the curator of the collection is also, in a sense, a manager. He is required to store his specimens in such a manner that he can readily retrieve them, for whatever purpose. As the manager, the curator must oversee all of the necessary functions to be certain that his institution serves the best purposes for which it was established and which it serves today. The number of different purposes which an herbarium and its contained specimens serve is essentially limitless. Each day brings some new application of the specimens in an herbarium collection. There is no predicting today what values any one specimen or any one collection of specimens may have for some future work. About all we can predict is that somebody will find new uses for our materials that we have been so cautious, so careful, to maintain.

In an ideal herbarium every specimen would be well identified, with no uncertainty. The realities, however, are so different that we all hesitate to expose our collections to anybody but our professional colleagues. However, if we consider that whatever information exists on the specimen, even though it be written in some very difficult script, is a means to identify the specimen, we are able to use the specimens for various purposes. The older herbarium specimens clearly are the least useful for ecological data; they also may have very poor geographic data. The many variations in kinds of data, and means by which the data were recorded (or omitted) on the specimens is so familiar to this audience that it needs no repetition. Nevertheless, with whatever variations that exist, we do generally manage to find the necessary specimens. If we cannot do so immediately, we usually correct mistakes at some later date and continue to improve the knowledge and information content of the herbarium.

If there were sufficient funds to assist the process of transferring all the data from the herbarium labels to some device which would use the many different facets of information on the label, which we know our

specimens could provide and which would aid the purposes not only of
taxonomy but of many allied matters, what could be done given the best
system available? Nobody in this audience needs be told how extremely
difficult it is to get additional funds for any purpose in any herbarium.
I believe, however, that by careful analysis of our problems and by
careful management techniques, we can indeed provide the herbarium
curator with some assistance from E. D. P. techniques, specifically,
information management systems. First of all, in management, we must
choose a set of priorities for our most important work: to which
activities we will devote most of our time; which of our collections
contain most valuable material; what parts of the world are most
significant for our individual herbarium functions? Priorities are
already established on these bases, and we spend most of our time
in providing certain kinds of information concerning these priorities.
The organisational priorities between herbaria, although informal,
are sufficiently well-defined for different herbaria to carry on major
functions without overlapping the work of others. For example, in
Britain, the priorities between the collections of Kew and those at the
British Museum (Natural History) have been established, and a
management decision made whereby certain different functions and
certain different kinds of collections would be emphasized by each of the
two institutions. Likewise, other herbaria know that Kew and the
British Museum (Natural History) provide these functions and therefore
do not try to duplicate those specified. These were formal management
decisions, and a set of priorities were established, indicating that
curators of collections are, indeed, managers.

As we move from the present methods of data recording into
techniques in the electronic age, we must exercise great care in order
to achieve the most effective, least costly procedures. At this point,
it would seem most efficient to employ persons whose specific training
is in the field of management science, since the functions and data of
herbaria are as complex as those of any organization in the world. In
management science students are trained to work in large organizations
to achieve the most efficient function in the most effective and least
costly manner. It seems reasonable to me to employ a management
scientist as a consultant at the beginning of our conversion to E.D.P.
methods.

Last in this discussion on management, is is necessary, for those
of you who have not devoted any effort to find out about E. D. P.
techniques, to have someone who can help sort out the many different
processes and means by which E. D. P. is accomplished. You have

heard here today, and in previous discussions, many different claims
for different E. D. P. methodologies. It is necessary to discover which
of these processes is most appropriate to the needs of the collections,
of the data, and of the objectives and priorities of the herbarium curator.
Again, the management scientist can be a great help in these functions.

2.   Description of TAXIR

The Taximetrics Laboratory, under my direction, has had a long
history of development of efficient computer-aided methods to assist
in various aspects of the taxonomic process. In 1966 we became aware
that one of the most impressive needs for the taxonomist was for some means
to handle most efficiently the enormous data loads, including the
specimens in herbaria. In 1967 we began to develop an information
retrieval system to incorporate data in any of the different forms in
which the data are expressed - either alphabetically, numerically, or
by combinations of alphabet and numbers; to organize these data to
retrieve any subset or combination; and to add, delete, or change any
data. We wanted the programs (software) to allow us to question the
data in the ways most meaningful to the taxonomist. In addition to the
requirement from the user's standpoint, which would include the ability
to use his own language and his own terminology as freely and as openly
as possible, we also required that the most efficient means of using
the computing machine would be a part of the design of our system. We
designed the means of storage and the means of retrieval of the
information to take advantage of the capacity of the computing machine to
carry out calculations, as described in one of our publications
(Estabrook and Brill, 1969). This is but one example of our efforts to
increase efficiency of use of the hardware.

I have summarized some of the major attributes of the TAXIR
system in an appendix to this paper. I have also included a very short
glossary which provides definitions of selected terms that are not
generally well known in the herbarium milieu.

a.   Using TAXIR for herbarium applications. Having described the
software package of TAXIR in a cursory manner, it is important to give
an example of the TAXIR system application. The most important
element to consider in the use of this, or any other computerised
information retrieval system, is the division of the sets of data which we
wish to capture into logical groups. We use a data bank to divide the
data into logical subsets. Even with the most powerful machines with the

largest memory units, it is necessary to limit the size of data banks to use the computing capacities at their greatest effectiveness and at the least cost. In large herbaria one could think of dividing data banks into two basic types: the one of most significance to this audience may be designated "curatorial" data banks, containing only the label information; the other type, containing data describing the plant material, does not concern us at the moment. Separate data banks may follow the logic already established in the herbaria: monocotyledons separately from the dicotyledons, and within these two major divisions, subdivision into appropriate size of data banks according to some classificatory scheme. For example, all of the palms or all of the grasses might form the basis of a single data bank.

In the TAXIR system it is possible to merge different data banks if needed. For example, if certain types of information were common to several data banks (such as collectors' names, localities, etc.), the TAXIR system can merge data from separate banks and put together the required information in a single data bank, to be used in the questions on the merged system. By dividing our data banks into logical sets we are always able to keep items in some manageable size and, by the same token, have available all the information upon call when necessary. Other kinds of data banks could be considered as required, either by the herbarium curator or by one of the taxonomists working on a specific problem in the herbarium. If one of the taxonomists, for example, specializes on one particular group of plants it would be possible for him to build his own data banks which could later be added to the general store of information for the whole institution. By using the TAXIR system in conjunction with all of the different functions of each of the curators (or visiting taxonomists) in the herbarium, we would be able to capture more rapidly the data from all of the herbarium. With careful management, data capture would proceed more rapidly than would be expected from a single calculation (i.e., it would take 100 man-years to capture all of the data from 4 million specimens at Kew). Such a prediction has very little value because we have not thought about the means by which we presently capture data and recognize that many workers are involved in a collective effort.

After consideration of the structure of our data banks themselves, there are clearly additional types of requirements to be certain that we will proceed in the most orderly manner for the efficient use of whatever software package we may choose to operate in our data bank. One of the required features is a careful standardisation to be certain that all collections are uniformly treated. In this connection, the discussion by Dr. Cutbill of Cambridge, who has spent a considerable amount of time on the problem of standardisation of museum materials, is extremely meaningful. Careful analysis of the means Dr. Cutbill has designed will be beneficial to the curators of herbaria as they begin the process of deciding the types of data which should be included in the data banks.

Once again, I should emphasize that in the considerations that must be made here, we must clearly make a separation between the substantive content of the data that may be found on the labels of herbarium specimens and the structure of these data. By this I mean the following: we recognize that many of the labels on herbarium specimens are inadequate, that the names given on the specimens may be wrong, and that there is generally a great problem with interpretation of cryptic symbols, etc., but we still use whatever is presented to find things. Placing the information, no matter how faulty or sketchy, into a system using good management techniques permits use of the specimen and, at the same time, much more rapid correction and improvement. Again, considering the structure of the data, we use these types of information - incorrect though they may be - as the means to find any specimen. Corrections or additions to data on herbarium specimens can be made much more efficiently when all of the data have been placed into some orderly data bank with some orderly use of an information management system.

b.   Interchange Between Different Software Systems.   We have heard presentations describing several different software packages that are used in various parts of the world. I do not wish to say that my system is better or worse than any of these other systems. Each of the systems is designed with specific purposes in mind and with specific attempts to solve problems. It is necessary to discover which of these systems is most appropriate in any particular setting. However, we must be aware that it is possible to employ different software systems with different computing machines and still share the collective data in the various systems by means of careful conversion programs. Each system discussed has a special format design by which the data are put into the computing machine. If we know the format design of any one system, we can convert from that particular format to the format of another system such that the data then becomes "compatible". Conversion routines can be generalised to the extent that they run extremely rapidly and very efficiently and with very little cost. Eventually, after some application of differing systems, it will be possible to compare the systems on an objective basis and, after the comparisons have been made, then, perhaps, will be the time for the adoption of one system rather than another. But my thesis is that we should not at the moment restrict ourselves to any one system, but test several general systems, each institution using that software package most easily available to it. Eventually we shall discover the common needs by this means rather than by trying to demand that all people conform to the same software system at the beginning. It is much more efficient and much less costly to progress as we are at the moment with many different systems running. We will be in a more secure situation to decide which system accomplishes our needs after a period of experimentation. Information retrieval systems which you have heard described here are not over six or seven years old and most of them are much younger than that. At this stage in the development of the powers and capacities of the computing machines for the uses in information management we are still in a development phase. We have efficient systems, to be sure, but none of the systems can be called perfect. In this light, this is a scientific endeavour. We expect some exciting developments in terms of both computing capacity and in terms of software efficiencies. We believe that the herbarium contains the most valuable sets of information describing our environment that exist in the world. It is time to put the full information contained in the herbarium to work and this can only be accomplished using good management practices and efficient E.D.P. methods.

REFERENCE

Estabrook, G. & Brill, R. (1969). The theory of the TAXIR accessioner. J. Math. Biosci., 5: 327-340.

APPENDIX I

Some attributes of TAXIR

TAXIR is an information retrieval compiler. That is, with the sets of instructions included in the program, differing data banks may be input to the computing system and from these a specified information retrieval system designed by simple, common language instructions, and with associated vocabulary, is achieved.

TAXIR is designed in modular fashion, with several "add-on" sub-
routines for various requirements. Examples of add-on routines are;
plotter programs to make maps and graphs; report generators to print
columnar data with headings; editing routines; and various statistical
packages (at the moment, the statistical packages attached were
designed by the Institute of Behavioral Sciences at the University of
Colorado).

TAXIR is written in FORTRAN IV, the most common compiler language
found in the world. Because of the exclusive use of FORTRAN IV, it is
relatively easy to convert from computing machines of one manufacturer
to computing machines of another manufacturer (the only requirement being
that the machine possess a FORTRAN IV compiler). At present, TAXIR
is running on Control Data Corporation machines (6400 and 6600), on
International Business Machines (several 360 and 370 series) and on
UNIVAC 1106 and 1108. Because of the modular design, it is possible
to use smaller computing systems than those mentioned above, such as
the IBM 1130, or the IBM 360/20 models. The only requirement for the
smaller machines is that they have associated random-access
peripheral storage devices.

The TAXIR version on the University of Colorado CDC 6400 is adapted to
run on the inter-active, time-share system, from remote terminals.
It also runs on "batch" mode for long runs with extensive print-out to
take advantage of lower cost operations.

The TAXIR system achieves extraordinarily efficient storage and retrieval
capability by employing as a base of design set-theoretic functions which
guarantee the most efficient use of storage capacity of the computer, and
at the same time provide extremely rapid access to the stored information.

The TAXIR system alone occupies about 15 K 32 bit words of memory,
and sizeable data banks can be contained within 32 K memory. Large
data banks take advantage of drum, disc and/or tape peripheral memory
to extend the size of the banks almost indefinitely. The largest data
bank ever tested on TAXIR had 106,000 items, with 50 descriptors per
item. Time to retrieve a single item in answer to a complex question
from the bank with 106,000 items in storage required ca. 2 seconds of
central processor time.

Any modern language using the Latin alphabet may be used as input to
TAXIR with rapid, accurate translation to other languages possible.

TAXIR does not require a pre-determined thesaurus of terms. Because of the design of data input as descriptor/descriptor state, the descriptions of each item become the language by which the data bank may be questioned.

Items, descriptors and/or descriptor states may be added, deleted, or corrected with a single instruction that is contained in the TAXIR compiler.

Data banks may be merged or reformatted under control of the TAXIR compiler.

TAXIR will (within the next 3 months) be completely flow-charted and documented, and a user's manual available. Complete flow-charts and documentation, along with listings and the user's manual, will be available at cost. Since the system was built with public funds, the system is in the public domain. No request from a serious user will be turned down. It is requested that any user who adopts the system become a member of a "users' group", **and that any further** development of the system made by any user be shared with the developers of the system. (This is, of course, not enforceable by any means, but it is expected that the spirit of free scientific exchange will prevail.)

APPENDIX II

Definition of some commonly encountered terms in computerized information retrieval

For a much more **complete** glossary, see: Vocabulary for Information Processing, published by the American National Standards Institute, and/or CDP Review Manual. A Data Processing Handbook, R. A. MacGowan & R. Henderson (eds.). Auerbach Publishers, Princeton, N. J., 632 pp.

batch mode (contrast inter-active). When using the computer, program and data banks are submitted to the computer to be run at the convenience of the computer center. There are no possibilities to alter the program, nor to ask further questions during the computer processing of the data. Batch mode is much less expensive to use, and is most often employed when large or long print-outs are expected.

character In the computing milieu, a single letter or numerical symbol.

compiler A software package which converts a set of instructions
    (program) to machine language.

data bank A collection of items (a set) with associated descriptors
    and descriptor states (q.v.). The proper design of a data bank is
    very critical in cost-effective use of any computer-aided
    information retrieval system.

descriptor In TAXIR, a single basis for description of an item.
    Example: collector, collector's number, generic name, species
    name, country where collected, flower colour, date of
    collection, etc.

descriptor state In TAXIR, a series of non-overlapping (mutually
    exclusive) descriptive statements (values) under each descriptor
    for each item (q.v.). Examples are given in the table below:

Descriptor	Descriptor State
flower colour	red white blue **red-blue**
leaf length	10 cm. 11 cm. 15 cm.
collector	Smith, R. Smith, J. Rogers, D. Rogers, W.

Note that the combination descriptor/descriptor state conforms
to the same construction as genus (noun), species (adjective).

Note also that descriptor states may be alphabetic, numerical, or combinations of these two. In TAXIR, there is no limit to the number of either descriptors per item or descriptor states per descriptor. The length (number of letters or numbers) per descriptor state is presently set at 90, but may be lengthened if need be.

In TAXIR, descriptors may be names, coded, or ordered (from-to), to give complete flexibility in description of the items.

field With reference to the number of letters or numbers used in any descriptor state, the place on the punch card (paper or magnetic tape) where one places data.

    a. fixed-field. A pre-determined number of spaces allotted on the punched card (frequently used in coded data).

    b. free-field. Within limits, any number of spaces allotted on the punched card. This is a feature of the TAXIR system.

hardware (contrast software) All the physical components of a computing machine, including input and output devices, central processing unit, storage units, cathode ray tubes, etc.

input Any method of data preparation for computer manipulation and the machines that accept data in the computer.

inter-active Any set of software and hardware computer configurations which permits the user to ask questions and receive answers sequentially, without resubmitting his program and data each time a new question or direction is submitted.

item In TAXIR, any thing or concept which may be defined as a basis for description, e.g., a specimen or a taxon.

"K" (as in 15 K) Shorthand, or jargon, in the computer milieu, standing for 1,000. Thus, 15 K = 15,000. Refers generally to the size of memory in any computing machine.

on-line When a user employs a computer with a remote terminal or time-share capabilities, he is said to be "on-line".

output Any means of presentation of the results of computer manipulation of some input under control of the computer program may be "hard copy" (typical computer print-out), a display on a cathode ray tube, or microfilm.

program (see also software) A set of instructions that directs the function of the computing machine to accomplish some task or set of related tasks.

remote terminal A piece of hardware connected at some distance from the computer via telephone line or microwave transmission. A means by which a computer user may communicate with the computer without having to visit the computing center. There are many levels of complexity of remote terminals.

software (see also hardware) A generic term that includes all types of programs that direct the functions of a computing system.

time-share The more sophisticated computing systems provide means by which several users may have programs running nearly simultaneously in a single computer.

THE USE OF THE SELGEM SYSTEM IN SUPPORT OF SYSTEMATICS

J. F. Mello

National Museum of Natural History

Smithsonian Institution, Washington, U.S.A.

Summary

SELGEM is a set of computer programs which have been designed to process the data of systematics in a variety of ways. Emphasis has been placed in SELGEM development upon input and manipulation of the data, although a fully operational set of output and reformatting programs is in existence. Under many circumstances the SELGEM system can be used to streamline the process of data collection, reduce or eliminate repetitive work, increase the accuracy of data assembled, permit the output of a greater variety of documents, and allow for more flexible uses of the resulting data base. The system is operational on a number of machines at the present time and is available without charge to any user provided that it is not used for profit. Further information about the SELGEM system itself can be obtained by writing to its creators - Mr. Reginald Creighton or Mr. James Crockett, of the Information Systems Division, Smithsonian Institution, Washington, D.C., U.S.A., 20560. Further information about uses to which SELGEM has been put at the National Museum of Natural History can be obtained by writing to the author or to Dr. T. Gary Gautier, National Museum of Natural History, Washington, D.C. 20560.

Introduction

The SELGEM system is a series of computer programs first designed to assist in the storage and retrieval of the data of systematic

biology. The chief purpose of this paper is to outline the uses which
SELGEM serves and the promise which it offers to systematists. As
background for this topic some general observations on the potential
value of the computer in systematics should be made.

The computer makes it possible for systematists to utilise much
more of the available observational and historical data than is possible
with conventional means, but new scientific procedures and approaches
will have to evolve as part of the proper utilisation of these data. There
are four basic ways in which the computer can assist in the care of
systematics data:

1. Large volumes of data can be accepted and kept available.

2. Relationships between the various elements of stored data
   can be retained.

3. Manipulation of the stored data can be accomplished, either
   by reorganising them or by testing relationships amongst the
   data items.

4. Data output can be obtained in many useful formats, including
   charts and maps, and often at great savings in time.

It is important to point out that much of the work of data
manipulation which can be carried out by the computer could not
reasonably be done by hand. Equally importantly, the computer simply
extends the scientist's ability to examine data and derive information
from them so that the benefits of computer use do not derive directly from
the machine but only indirectly as scientists improve in their ability to
utilise larger volumes of information. It also should be recognised that,
along with the opportunity which the computer presents, there is also the
possibility of misuse. Discipline, in the form of standards for the data
which are to be entered and conventions which will govern how the data
are collected and organised, is necessary to assure that maximum
long-range advantage can be derived from the stored data.

Acknowledgements

The development of SELGEM has been a team effort involving many
participants. Chief credit must go to the programmers and systems
analysts at the Smithsonian's Information Systems Division, and
primarily to Mr. Reginald Creighton and Mr. James Crockett, for

creating the system. Mr. David Bridge of the National Museum of
Natural History contributed greatly to the conceptualising which
preceded much of SELGEM development, and he continues to provide
much of the management skill which has made the system come alive
at the National Museum of Natural History. Credit must go to Dr.
Richard S. Cowan, who had the vision to recognise the value of the
computer in systematics and established the Automatic Data Processing
Program at the Museum, and to Dr. T. Gary Gautier who has provided
leadership for the program since mid-1973.

The data of systematics

The basic data of systematics are observations on the identity,
morphology, geographical and environmental distributions and
relationships amongst species and specimens of plants and animals.
The process of data collection is carried out in the field and in the
laboratory and is usually accompanied by the collection of specimens
which document the observations. The usually large body of data
collected during these generative stages of investigation is digested
for purposes of publication, along with the derived information or
conclusions developed by the systematist. Some of the basic observational
data and some of the derived data are commonly recorded on the labels
associated with the specimens before they are entered into museum
collections.

It is important to note that the two usual means by which information
is made available to future generations of systematists are through the
publication of the derived conclusions, along with some portion of the
original raw data, and through the preservation of certain elements of
information on labels associated with the specimens collected. Field
notebooks and other informal documents which contain the original
observations are generally not widely available to other systematists
and are not incorporated into the general corpus of knowledge. A great
opportunity exists, for the individual systematist and for systematics
as a science, by the use of the computer to receive, assemble,
manipulate and output the basic data collected in the field and in the
laboratory. If the basic precaution of careful data assembly is followed
at the time of data collection then it should be possible for the
individual researcher to use the assembled data for his own purposes
and also to pass on the basic information to his successors in a much
more complete and readily retrievable state than is customary or
possible today.

Specimen-related data

For most of the specimens stored in the world's museums of
natural history the only data available are those which are contained
on their labels and those which have been published about them. In
general the published information is largely derivative, with little of
the original 'raw' data included. Thus the specimens and associated
label data must largely stand on their own as basic documentation of
nature. A great deal of valuable information can be derived from
these labels, and the long-range effect of accumulation of these data
into computer files can be a substantial improvement in the usefulness
of the collections themselves. There can also be beneficial short-range
effects through use of the computer for data assembly, verification and
replication. However, basic to the successful use of the computer and
the storage and retrieval of specimen-related data is the establishment
of and adherence to standards of data organization.

The need for standards

The vast amount of data available for specimens already in hand,
and the quantities of new data gathered each year as more specimens
are collected, makes it necessary that there be consistency in data
assembly. While the computer can provide access to large volumes of
data, these data must be internally consistent if maximum benefit is
to be derived. The establishment of rigid formats within which the
data are to be collected is not necessary, although it may be convenient
to use such formats for particular collections. Depending upon the
computer software system used, and perhaps upon the computer as well,
there may be conventions which must be followed in indexing records in
order to properly tag them for internal computer record keeping. Use
of formats and adherence to computer conventions are of little importance,
however, as compared to the importance of establishing standards for
the ways in which the data themselves will be recorded. It is necessary
that where the same data are to be recorded for a group of specimens
they be recorded in the same ways for each specimen. Thus, for
example, the use of a geographical term must mean precisely the same
thing for all specimens to which it applies in any particular data base, or
else the precision with which the specimens can be retrieved in terms of
geography will be reduced.

As the computer begins to be used extensively for the storage and
retrieval of information about systematic collections, the various
scientific societies should take it upon themselves to establish

standards which will be used consistently in the recording of information about the organisms of concern to them.   These professional groups should also communicate with each other about standards which they intend to establish so as to achieve the highest possible degree of compatibility amongst as many disciplines as possible.   In the United States the Association of Systematic Collections is making good headway towards this kind of cooperation.   However, there will be many data classes which cannot conveniently be standardised, and there may be unusual circumstances which make it impossible to follow standards in treating a particular collection.   The data assembled under these conditions can still be very useful to systematists, providing it is recognised that they are not completely compatible with standards and thus must be treated most carefully by their users.

Possible future benefits of computer use in systematics

Morphological descriptors can be stored, processed and retrieved through the use of the computer, and the potential value of the kinds of comparisons which can be made would seem to be great indeed.   However, the computer also has the potential of either storing directly, or referencing, visual images of morphological elements.   These images would be largely self-descriptive and could eliminate the need to establish and maintain a glossary of written descriptors for the various groups of organisms.

The ability of the computer to handle large numbers of variables, for each of which many separate pieces of data may be recorded, offers systematists the opportunity to record more information and to make better use of that information.   Numerical taxonomists have already begun to take advantage of this capability.

The increasing cost of publication of systematics data, and the increasing volume of information being generated in systematics, is severely pinching many journals.   The computer can be used to output some or all of its stored data in page format, and thus can be used as a short cut in the publication process.   Direct offprint from computer copy is not very attractive, but techniques exist for processing computer tapes through photo-composition equipment to create much more acceptable copy, again at great potential savings.   Use of the computer in the publication process could become simply a final stage in its use in the assembly and manipulation of the data for original scientific purposes, and in turn would help to stimulate the use of the computer in the basic research aspects of systematics.

The computer is currently being used to store and retrieve the data of systematics both at the research level and for documentation of museum specimens. It will continue to be used in both of these ways but its use can be made much more effective if the data are initially recorded in machine-readable form either in the field or in the laboratory, or when specimens are accessioned into collections. The value of the computer as a long-term reservoir of complete or nearly complete basic information will probably be great if the data are collected under control and in adherence to standards. Strong leadership and close participation will be needed among those who are responsible for organizing systematics data as well as between the scientists who assemble the data.

Development of the SELGEM system

In 1967 the Department of Health, Education and Welfare, Office of Education, awarded a three-year grant to the Smithsonian Institution for joint research by the Information Systems Division and the United States National Museum of Natural History on the feasibility of computer use in the storage and retrieval of data about museum specimens. Out of this work came the pilot Natural History Information Retrieval (NHIR) System which consisted of a series of computer programs that permitted the storage and retrieval of specimen data. The design of the system included the construction of a code number for each specimen recorded, and also called for the verification of geographical data and the organisation of taxonomic data in accordance with previously established classifications. Unfortunately, costs for entry of data into the NHIR system were quite high and many of the capabilities of the system were inefficient at low data volumes. Therefore it was necessary to set new criteria for operation at a more general and basic level and to develop a largely new set of programs which were less specific and therefore more flexible.

The SELGEM system was developed as a better alternative, building upon the programs and experience acquired from the creation of the NHIR system, to meet the more practical needs of collection management at the National Museum of Natural History. Some of the basic initial requirements for design of the SELGEM system were that:

1, natural English language could be used in the recording of data - data coding was not to be required,

2, input requirements were to be very flexible in order to permit a wide variety of input formats and the use of several kinds of input techniques (paper tape, magnetic tape, optical character recognition, punched cards),

3, arranging of the data for input and also for output was to be entirely flexible and controllable by the user,

4, editing routines should be provided for maximum assistance to those creating the data bases, and for good error control.

In order to provide the degrees of freedom for data entry and output described above, the SELGEM system was designed as a tag system. A tag is a three-digit number which identifies a category of data. For example, genus is a category of data, and individual generic names are the data elements which are placed in that category. A three-digit number is assigned as a tag to be used by the computer in place of the category name 'genus' for purposes of efficiency in internal computer processing.

In establishing any data file in the SELGEM system it is first necessary to examine the structure of the data to be processed in order to determine the logical categories of data which are to be recorded. Each of the selected categories is then clearly defined as to the data format which will be used in recording the data, the spelling and abbreviation conventions which will be followed, and the standards which will govern the meanings of the descriptive terms used. Once this process of logical analysis and definition has been completed and the data categories and three-digit tags have been established, the process of input can begin.

Key aspects of SELGEM use

Not all of the capabilities of the SELGEM system can be explained in a short paper, and certainly no good idea can be given of the variety of ways in which SELGEM is used to serve the many individual projects in which it is employed. Nevertheless it is possible to give some idea of the uses to which it is put by means of brief descriptions and examples.

Input strategies

Whenever possible, data are taken directly from specimen labels in order to avoid the extra work of transcription onto forms. However,

in cases where the specimens cannot conveniently be brought to the
data input machines, it is necessary to transcribe the data onto
intermediate data-capture forms. The programmable punched-paper-
tape typewriter is used extensively for data input at the National
Museum of Natural History and is particularly well suited for the
purpose because it can be easily programmed to automatically retype
largely repetitive data and to assist the operator by automatically
typing cues when new data elements are to be entered.

Frequently when dealing with museum specimens there are elements
of data held in common for long series of specimens. For example,
10 or 20 or more specimens might have been taken from the same
geographical locality by the same collector on the same day. While
it is necessary that each specimen has its own label - and traditionally
these would be created by writing or typing all of the data on each label -
it is possible to gather the information held in common just one time
through use of the SELGEM system and then to use the computer to
create labels containing all of the information for every specimen. This
'header-trailer' capability of SELGEM permits great savings in time,
makes possible considerably increased accuracy, and eliminates a dull
and demeaning activity from the tasks of collection managers. Header
information is that information which is common to a series of
specimens, and it is recorded as a separate header record which will
precede the unique information descriptive of the individual specimens
in the series. Trailer information is information unique to the
individual specimen. A header record and its associated trailer
records can either be preceded or followed by regular records without
causing any problems as far as the SELGEM system is concerned.

The SELGEM system can accept data created through optical
character recognition (OCR). The chief advantage of this method is that
a fairly inexpensive and widely available instrument - an electric
typewriter with appropriate type font - can be used to prepare machine-
readable copy. An optical scanner must be used to "read" the typed
sheets and duplicate the data onto magnetic tape, which is then
processable by a conventional computer.

Processing techniques

When new data are submitted to the SELGEM system for initial
processing an update report is printed which lists all of the data entered.
This report is carefully examined for errors, which are usually
corrected by means of punched cards. Once the new data are clean

they can be entered into the master file, which is the chief archive
to which additions are made and into which corrections are entered.
A master list may then be printed to provide an eye-readable copy
of the computerised data for future reference.

SELGEM can be of significant help to editors because it can be
used to prepare special listings which speed up and increase the
accuracy of error checking. Through its category- and data-frequency
report routine SELGEM can put the data which it finds in any data
category (such as genus) in alphabetical or numerical order. Duplicate
expressions can be eliminated from the listing so that each generic name
appears only one time, and the numbers of all specimens to which that
name has been applied can be listed beneath the generic name. By
scanning the list of unique names the proof-reader can easily identify
errors in spelling, and corrections can be made by punching a card with
the identifying number and the corrected name for each specimen
bearing the incorrect name. Use is commonly made of this capability
in editing other taxonomic fields as well, and also for treating
geographic, collector and sample information. The human editor need
only concentrate upon any particular taxonomic, geographic or other term
a single time, satisfy himself that it is correctly used or make
corrections where necessary, and then be assured that wherever that
term appears in the file it will be exactly as he has seen it on that
single occasion. This is a particularly valuable capability when data
which might be deeply embedded in context, and which thus might be
very hard to correctly and consistently proof by conventional methods,
must be carefully edited.

SELGEM output

Data stored in the SELGEM system can be output in a variety of
ways. In addition to conventional output from high-speed printers
there is the creation of graphs and charts through use of plotters,
creation of computer-output microfilm, and output onto magnetic tape.
At the National Museum of Natural History the high-speed printer is
used as the chief means of SELGEM output. The system is used
continually to create documents needed for collection management such
as catalogue listings and specimen labels, but it can also be used for
more sophisticated types of output.

The SELGEM output programs have considerable flexibility in
reformatting the data. This means that data items which may have

been entered first during input can be placed in the middle or at the end
of a record on output. Also, the horizontal positioning of any data
category, and the spacing between data categories, can be controlled.
And data categories which may be included in original records but
which are not wanted on output can be deleted.

While SELGEM output capabilities have been used most extensively
to create specimen labels, they have also been used to create composite
listings for collections (Collier, 1971; anonymous, 1971; Shetler, 1973).
As data assembled for additional collections is completed, new catalogues
will be prepared through the use of SELGEM and its output capabilities.

Merging of data in SELGEM

The header-trailer method of data input described earlier permits
the user to enter, just once, those general classes of information
pertaining to a series of specimens. SELGEM can expand each record to
contain all data fields during processing. In addition to this capability
for expanding data files, SELGEM can be used to create more comprehensive
files by assembling data which are introduced in two or more separate
series of data inputs. For example, it is possible to record routine
catalogue information such as taxonomic name, collector and geographical
location at one time, and then to supplement the data file with
morphological observations or names of associated specimens at a later
time. Each set of data may have been assembled for valid immediate
purposes, such as the creation of specimen labels or the analysis of
morphological characteristics, but the merged data file might well be of
value for other purposes.

Additional products

Although SELGEM was designed and is chiefly used for the processing
of data about specimens to collections it has also been used to produce other
products not directly related to collections. Data about the taxonomy,
geographical distribution and environmental adaptations for the Hymenoptera
(ants, bees, wasps, etc.) are being entered into SELGEM (Krombein,
Mello and Crockett, in press). The purpose in creating this file is to use
the computer as an aid in assembly and editing of this very large volume
of material and then finally to output clean tapes which can be used with
a computer-driven linotron. This device will produce a new catalogue
of Hymenoptera in a variety of type-styles and fonts. SELGEM, through
its editing capabilities, makes it possible to ensure that the data are as

clean as possible, and the amount of human labour - especially the labour of professional entomologists - should be considerably reduced. Furthermore, the data base built during this cycle will be maintained in machine-readable form and will be available for use 5 to 10 years hence when it will be necessary to create the next catalogue.

SELGEM is also being used to output magnetic tape which can be used in automatic plotters. The charts, graphs and maps which can easily be produced by these machines are of great value in analysing relationships amongst taxa or amongst specimens under study.

SELGEM can also be used to reorganise data so as to interface with statistical packages or with other data storage and retrieval programs. Further information about the capabilities of the system are available in Creighton & Crockett, 1971; Creighton et al., 1972.

Technical information about SELGEM

1. COBOL (Common Business Oriented Language) is one of the most widely used programming languages for computers and has been used in the writing of all SELGEM programs.

2. A Honeywell 2015 computer is in use at the Smithsonian for the operation of SELGEM, but the system has also been made operational on the following computers at over 15 other institutions: IBM 360, CDC 3100, CDC 6400, UNIVAC 1110, GE 635, and also Burroughs and ICL computers.

3. The system requires a minimum computer configuration of 4 tape-drives, 20,000 characters of core and a COBOL compiler.

4. Installation of SELGEM at any particular computer centre can cost from as little as $500 for transfer to another Honeywell 2015 to as much as $5,000 (including personnel costs) for transfer to a different machine. Unfortunately it is not possible to guarantee any particular figure for the cost of establishment of SELGEM (or any other computer system) since there will be differences in the skills of those who are assigned the task of making the programs operational and also differences in each particular computer environment.

5. A single programmer/analyst should be able to maintain the SELGEM system once it is operational, providing that there are not too many active files which might require minor program modifications. After production of a small group of files is smoothly organized it is likely that a trained museum technician can take over from the programmer/ analyst.

6. The prorated cost for <u>computer time</u> to build a file and obtain various information products is currently about 23¢ per record at the National Museum of Natural History. Input costs account for about 17¢ per record, and about 6¢ per record goes for editing.

7. SELGEM can be obtained by any museum or educational institution anywhere for only the small cost of duplicating the programs and documentation. No fees are charged for the use of SELGEM and no agreements are necessary except that there is a prohibition against reissuance, resale or use of SELGEM for profit. SELGEM users are kept up-to-date on new developments through a newsletter entitled MESH, which also provides them a means of learning about other users. All SELGEM users are encouraged to exchange ideas and programs.

Conclusion

Complete information about the capabilities of SELGEM cannot be provided here but the following summary will give an idea of the system's capabilities. The SELGEM system consists of programs which:

a) accept data on input; b) format or reformat data after input;
c) edit the data and prepare listings; d) absorb corrections and prepare correction lists; e) write the data onto magnetic tape;
f) retrieve the data in a variety of different formats; and g) output the data on paper, punched cards, computer-output microfilm, or magnetic tape.

There are presently over 50 separate SELGEM data files in use at the National Museum of Natural History, and all seven scientific departments of the museum are being served. Emphasis has chiefly been placed on the recording of information about incoming specimens, type specimens, or collections of particular importance because of their uniqueness or current research interests.

The introduction of computer data storage and retrieval into the National Museum of Natural History has been accompanied by human reactions. It would be unfair as well as uninformative to characterise the response to computerisation as either favourable or unfavourable. Scientists have been ready to accept the computer in the museum, provided it could be shown to be an effective tool of benefit to the legitimate pursuits of systematics. The strongest logical argument for computer use in the museum environment is the responsibility which systematists and museologists have to their fellow scientists for care of the specimens, and the vital accompanying data, which serve to document systematics. There are some who feel that the money spent on improved collection management and information interchange through computer use might be better spent in support of more research. However, museums must also look to the discharge of their responsibility to care for what has already been collected as well as to provide support for future investigations. A balance must be struck between support of new research and management of information gathered during past and current research. The emphasis is now, and must always remain, upon support of new research, but not to the exclusion or neglect of the continuing responsibility to care for what is already collected. Through better care of the information which forms the history of systematics it should be possible for scientists to conduct more research with greater certainty of the validity of the data with which they deal and with less time spent in resurrecting old data.

While it may cost as much as an additional $2.00 per specimen, when all costs are considered, to enter specimen information into a computer system, this expenditure is minor compared to the $50 to $100 per specimen which may have already been invested to collect, prepare, identify, and preserve the specimen.

REFERENCES

ANONYMOUS (1971). Fossil Marine Mammals in the Collections of the Department of Paleobiology, United States National Museum. Unpublished, 173 pp. plus appendices.

COLLIER, F.J. (1971). Catalogue of type specimens of invertebrate fossils: Conodonta. Smithsonian Contributions to Paleobiology, 9: 256 pp.

CREIGHTON, REGINALD & CROCKETT, J.J. (1971). SELGEM: a
    system for collection management. Smithsonian Institution
    Information Systems Innovations, 2(3): 24pp.

CREIGHTON, REGINALD; PACKARD, PENELOPE; & LINN, HOLLEY
    (1972). SELGEM retrieval. Procedures in Computer Sciences
    (Smithsonian Institution), 1(1): 38 pp.

SHETLER, S. G. (1973) An introduction to the botanical type specimens
    register. Smithsonian Contributions to Botany, 12: 186 pp.

DISCUSSION, AFTERNOON, THURSDAY, 4 OCTOBER

This was a general discussion, led by Mr. B. J. Harwood, of the papers read during the afternoon.

<u>Mr. B. J. Harwood</u>. A system is wanted which will provide taxonomists with the information they require. It is impossible to have a system which can provide all the information for all requirements; therefore we will have one with some constraints. The standards will be the key words, descriptors and definition of the record. The system must be relatively simple in order that it may be understood by the console operator and by the systems analyst who will be required to put right any faults that may develop. It must be easy to maintain, modify and amend. The cost benefit must be in proportion to its usefulness. An international system will have both a language barrier in verbal communication and in the programs. The programs must be compatible between terminals of different host computers in order that there can be a full interchange of data, for example, by magnetic tapes. I wonder whether the system should be user-oriented, so that the user can define what the system should be capable of, or whether this should be the responsibility of the systems analyst? How easy is it to maintain SELGEM and TAXIR? How long does it take to understand either system?

<u>Dr. D. J. Rogers</u>. TAXIR is no more complex than any comparable system. It is impossible to answer the second question because it depends upon so many variables.

<u>Dr. J. F. Mello</u>. Our Museum provides the Information Systems Division, the computer operations unit of the Smithsonian Institution, with approximately $70,000 per year of support for programming assistance. Programming support for SELGEM, whether derived from the Museum of Natural History or provided out of central Smithsonian funding, is planned to continue for the indefinite future. Thus maintenance of the existing system and development of new programs to further enhance the system seem assured. All SELGEM programs are within the COBOL language and are not difficult to make operational on most computers. The programs can be obtained without charge except for costs of duplication and the price of magnetic tapes or other transfer media.

<u>Dr. J. L. Cutbill</u>. It takes us at the Department of Geology at Cambridge 3 months per year maintaining the programs (getting the bugs out).

Dr. S.G. Shetler. We should look at the character of the available systems and assess them as to how applicable they are to our requirements. These systems must be closely linked to the means of providing the data for them.

Professor C.H. Oppenheimer. With our Environmental Data System (ENVIR, 64 system) it is possible to learn to use it in a few weeks. Students with some knowledge of computers can use it easily and we have been able to teach visitors to use it very quickly because the system is easily understood - in the English language. It is very versatile in obtaining different types of information.

Mr. J.P.M. Brenan. Are these problems concerned with the interchange of information between TAXIR and SELGEM systems?

Dr. D.J. Rogers. Each system has a different structure but there is no problem in writing a conversion program to overcome this difficulty.

Dr. D.B. Williams. How does the system handle data which contains groups of repeated entries, for example a reference with author, date and title or identifications and identifier?

Dr. J.F. Mello. SELGEM provides the ability to handle hierarchical data provided that the data are themselves consistent in their organisation. Treatment of repeated data fields under more general headings, provided consistency is maintained, is not hard to accomplish.

Dr. D.J. Rogers. One way of dealing with this problem is to make as many data banks as needed. A merging command deals with this.

THE APPLICATION OF ELECTRONIC DATA-PROCESSING TO THE MAPPING OF PLANT DISTRIBUTIONS

James H. Soper

National Herbarium of Canada

National Museum of Natural Sciences, Ottawa, Canada

Summary

The first part of this paper outlines the important features of distribution maps as used in botany, describes how they are constructed and discusses the introduction of machine-mapping methods in the early 1960's. Examples of hand-plotted and machine-plotted maps are given. The main techniques for the automation of mapping plant ranges are described as well as the special capabilities which computers provide for mapping. A comparison is made of the different kinds of machines which can be used for machine-mapping from the standpoint of Input, Output, Control, Operation, Restrictions, Advantages and Disadvantages.

The second part of the paper describes the development of mapping programs for plotting the distribution of the vascular plants of Southern Ontario and examples are shown of output which can be used directly for publication. Reference is made to E.D.P. systems tested at the National Herbarium of Canada and to the current project to develop an information-retrieval system linking label production and automated mapping to the formation of a data bank of botanical distribution records. The importance of improving the quality of the data on future herbarium specimen labels is stressed and attention is called to the value of local gazetteers based on collections in herbaria.

A list of selected references on machine-mapping and related fields is provided.

Part I

Introduction

Definitions. Distribution maps are extremely useful devices for indicating where within a given area a particular organism occurs or has been found in the past. They are used extensively in plant systematics, in floristics and in plant geography. For example, current monographs often include a map of the range of each major taxon studied. State or provincial floras and other local floristic studies frequently contain distribution maps and in some cases these maps cover a much larger area than that of the region under study. A map may show the distribution of a species for the entire world, for a hemisphere, a continent, a political division or any smaller area. A chart quadrat as used in field studies in ecology is essentially a composite distribution map of all the species growing on a portion of the earth's surface and the area involved may be as small as a square metre. One of the most interesting aspects of phytogeography is the study of the patterns of distribution of plant species in relation to current or changing environmental factors and historical events. Accurate distribution maps provide the material from which we can formulate hypotheses to account for the occurrence and migration of species and for the development and composition of the flora.

How maps are made. In order to make a distribution map one needs the two obviously essential ingredients: (1) an appropriate base map and (2) data in the form of distribution records. In selecting a base map one should consider first the area being covered by the study in relation to the total natural area of the taxon being mapped since this may affect the choice of the most suitable type of projection. Other considerations are the size of the map and the scale in relation to the amount of reduction required if it is to be used for publication, and the need to include certain physiographic features and political or other boundaries. Finally, a decision must be made on the kinds of symbols (size and shape), lines or cross-hatchings which will display the information to best advantage.

The data may consist of only one kind of distribution record such as specimens which have been examined during the course of monographic study. It usually happens, however, that a more detailed picture of the known range of a species can be obtained by using additional information such as sight records (observations made in the field when no specimen was preserved), published reports and various other records such as data from unpublished sources, correspondence and specimens submitted for identification but not suitable for preservation (see example in Fig. 1).

Until comparatively recently, all distribution maps were made by hand, a slow and tedious procedure which usually requires the preparation first of a work copy and then a final inked copy suitable for reproduction and printing at a reduced size.

Lettering templates and special mapping templates are helpful in the production of a neat "dot-map" with a clear legend. However, the value of the map will depend chiefly on the reliability of the basic data and the accuracy with which the locations have been plotted on the particular base map in use.

Machine-mapping

The New Era. The Atlas of the British Flora which appeared in 1962 marked the opening of a new era in the automation of mapping biogeographical data. That publication contains some 1700 distribution maps, the result of data accumulated over a ten-year period and coded onto small punched cards, sorted and printed on standard base maps by means of data-processing equipment (Fig. 2). Actually, no computer was involved in that operation and it is therefore more appropriate to speak of machine-mapping than computer-mapping. Although machine-mapping may involve the use of a computer, it may also be carried out entirely on associated (peripheral) equipment such as tabulators, high-speed line printers or mechanical plotters controlled by punched paper tape or magnetic tape.

The Flora of Warwickshire project, begun in 1950, adopted a computer technique for the storage of data and ultimate conversion to a paper tape format for input to a mechanical plotter. The result was a truly "Computer-mapped Flora" (Cadbury et al., 1971). The computer checked the format of the distribution records, sorted and counted the records by unit grid areas, recorded the habitat types, calculated frequencies and generated the instructions for the plotter to draw the appropriate symbols in the corresponding positions (grid squares) on the final output map. Printing for publication was done using a blue outline for the county and its grid, and a separate printing in black of the

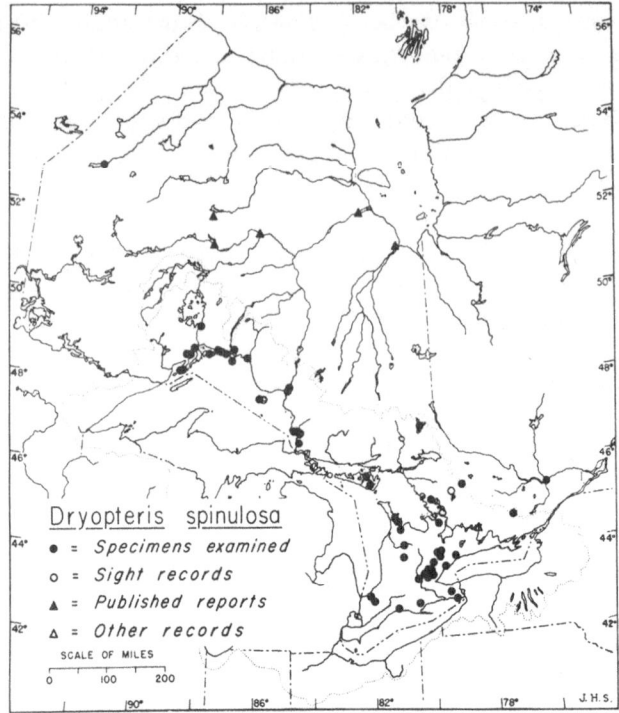

Figure 1

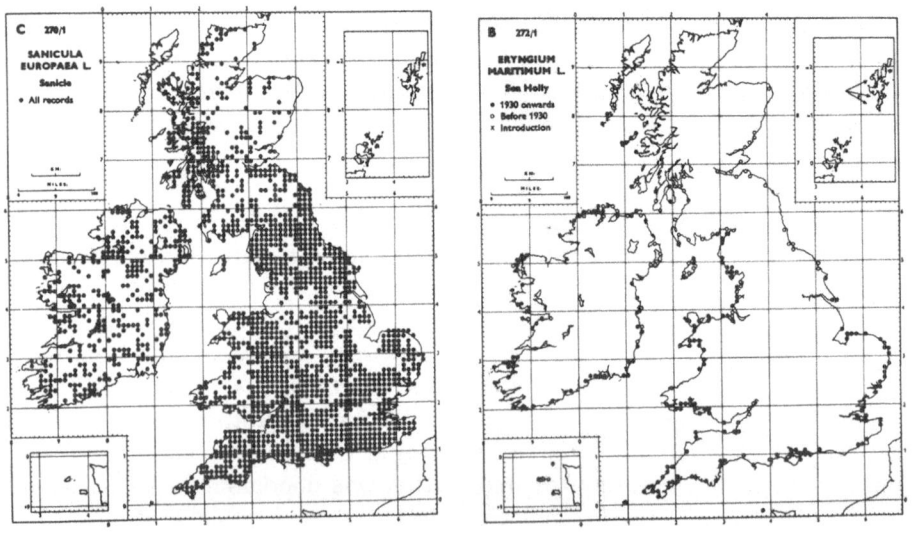

Figure 2

distribution data using symbols to indicate habitat types and species frequencies (Fig. 3).

Discussion of techniques. Depending on the aims of the project, there is a choice of several different combinations of techniques which may be used to automate the production of distribution maps. A simple approach is to plot symbols on plain paper and add the base map by means of an overlay drawn on a sheet of transparent plastic. A second method is to plot the symbols directly on an appropriate base map. Perhaps the most sophisticated procedure is to draw the map and plot the symbols in the same operation. Two indirect plotting techniques are the photographing of maps displayed on a cathode ray tube screen and the projection of maps on to photographic film for subsequent printing. Some of these operations can be performed on more than one kind of equipment, but others are restricted to a particular kind of machine. An attempt will be made to examine the main types of equipment available for machine-mapping from the standpoint of comparing Input, Output, Control, Operation, Restrictions, Advantages and Disadvantages (see Tables 1 to 7).

Special capabilities which a computer provides for mapping

1. To store several base maps and select for special job or use one part of map for specific area.

2. To change the scale and transform the projection, with accompanying transformation of data to corresponding true plotting positions.

3. To sort records to eliminate duplicates and/or establish priority for selection of important categories.

4. To convert from Universal Transverse Mercator to Latitude/ Longitude (and vice versa) and from either of these to X-Y plot position.

5. To store a gazetteer and calculate plotting position of locations having modifiers.

6. To smooth outlines for large-scale maps, deleting small islands, extra rivers, secondary boundaries, etc.

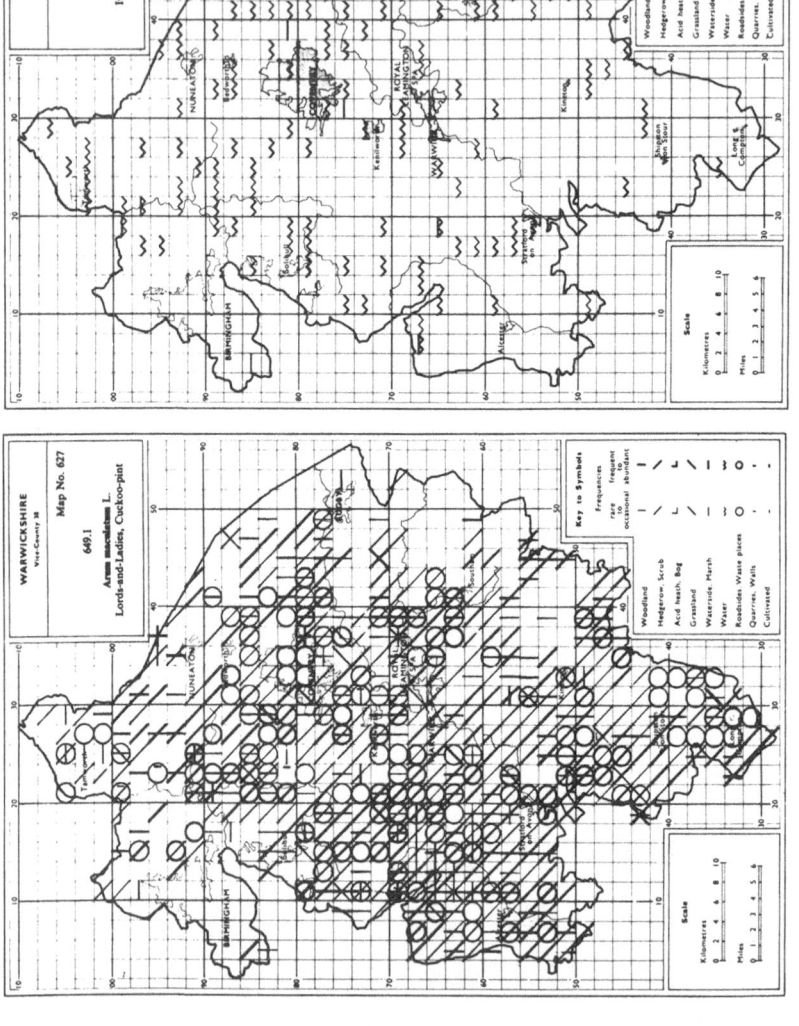

Figure 3

Table 1 - THE TABULATOR

Input	Punched cards.
Output	Continuous forms, plain or preprinted.
Control	A wired panel ("plug-board") and a paper tape belt (for line and skip control).
Operation	Off-line. Prints one line at a time.
Restrictions	Cards must be presorted in axis parallel to direction of paper movement. Normal spacing is 10 characters per inch and 6 or 8 lines per inch, but a modification for 10 lines per inch can be installed. May require mounting special paper.
Advantages	No computer time required. Preparation of input requires only a card punch and a punched card sorter.
Disadvantages	Cannot draw the map. Produces only grid maps with limited selection of characters. Slow operation. Presorting necessary. Grid must conform to a map co-ordinate system or requires a gazetteer of artificially coded locations corresponding to actual locations.

Table 2 - TYPEWRITER CONSOLE OF A COMPUTER

Input	Manual (keyboard of console) or via any computer input medium (cards, tape, disk, etc.).
Output	On continuous forms, plain or preprinted.
Control	By a program in computer memory.
Operation	On-line. Prints one character at a time.
Restrictions	Requires on-line time (C.P.U.) Spacing is usually 10 characters per inch and 6 lines per inch. May require mounting special paper.
Advantages	Suitable for small forms. Width of 80 to 132 characters.
Disadvantages	Expensive and prevents normal computer control signals. Comparatively slow since it is a character-printer. Cannot draw a map. Maps are crude due to spacing restrictions and use of strings of symbols to generate outline.

Table 3 - HIGH-SPEED LINE PRINTER

Input	Magnetic tape or punched card deck.
Output	On continuous forms, plain or preprinted.
Control	Magnetic tape or program in computer memory.
Operation	Off-line or on-line.
Restrictions	Spacing usually 10 characters per inch and 6 lines per inch. May require mounting special paper. Limited to available symbols on print chain and to 132 print positions per line.
Advantages	Rapid printing (300 - 500 lines per minute). Can produce range of density symbols by overprinting several characters.
Disadvantages	Expensive if used on-line. Cannot draw the map. Maps are crude due to spacing restrictions and use of strings of symbols to generate outline.

Table 4 – THE SYMBOL PLOTTER

Input	Manual, punched cards, paper tape, magnetic tape.
Output	Continuous forms, plain or preprinted (Drum plotter) or single sheets, plain or preprinted (Flat-bed plotter).
Control	Simple digital commands or program on magnetic tape or in computer memory.
Operation	Usually off-line but may be controlled by a computer.
Restrictions	Preprinted forms must be carefully aligned before plotting begins.
Advantages	Can produce "dot maps" with a variety of available or custom-made symbols. Can use several colours. Can print legend.
Disadvantages	When using single sheets with a preprinted base map, each map must be carefully aligned before plotting begins, i. e. an automated series is impossible. Even when using preprinted maps on a continuous form with automatic paper advance, variations in humidity during storage of paper or between different plotting sessions may change the dimensions of the paper and this can result in offsets of symbols which build up during a run (cumulative error).

Table 5 - THE LINE-DRAWING PLOTTER

Input	Manual, punched cards, paper tape, magnetic tape.
Output	Continuous forms, plain or preprinted (Drum plotter) or single sheets, plain or preprinted (Flat-bed plotter).
Control	Simple digital commands or program on magnetic tape or in computer memory.
Operation	Usually off-line but may be controlled by a computer.
Restrictions	Preprinted forms must be carefully aligned before plotting begins.
Advantages	Can produce "dot maps" with an infinite variety of symbols including those available with software package and any others added by subroutines. Can vary the scale, use several colours, print legend.
Disadvantages	When using single sheets with preprinted base map (Flat-bed plotter) each map must be carefully positioned before plotting begins. When using continuous preprinted forms on a drum plotter, changes in humidity can affect dimensions of the paper and result in offset errors in the plot.

Table 6 - CATHODE-RAY-TUBE DISPLAY (C.R.T.)

Input   Manual (keyboard of console) or via magnetic
                         tape or computer program.

Output   Picture on C.R.T. screen which can be viewed
                         and photographed.

Control   Via console (keyboard) or program in
                         computer memory.

Operation   On-line.

Restrictions   Requires C.P.U. time.
                         Dimensions limited by effective screen size.

Advantages   Rapid display   (immediately available).
                         Can examine map for errors before
                         photographing.
                         Map can be stored and manipulated or up dated.
                         Map can be transmitted to another C.R.T.
                         screen.

Disadvantages   Expensive.
                         Photographs are not entirely satisfactory for
                         direct publication.

Table 7 - THE PHOTO-PLOTTER

Input	Magnetic tape
Output	On negative film.
Control	Program in computer memory.
Operation	On-line or off-line.
Restrictions	Needs C. P. U. time or separate control unit.
Advantages	Can vary the scale. Can store data for rapid updating.
Disadvantages	Expensive. Cannot see the map until the negative is developed or print made.

Part II

Report on the development of mapping and labelling programs

1.   University of Toronto. Apparently the first experiments in machine-
mapping of botanical data to be carried out in North America were those
undertaken at the University of Toronto in 1963 when a Friden Flexowriter
was installed in the Herbarium of Vascular Plants (TRT) to capture
phytogeographic data in machine-readable form (Soper, 1964). The first
step involved the production of catalogue records and a punched paper
tape. The paper tape was converted into a series of 80-column punched
cards, each card containing an extract of the essential data from the
corresponding catalogue entry. Later, the program was modified to
include the production of herbarium labels from the paper tape and
punched cards were generated directly on an I. B. M. Card Punch
attached to the Flexowriter.

In May 1963, the first automated maps were produced by plotting
the individual records on gridded paper by means of an I. B. M. 407
Accounting Machine and adding the base map as a transparent overlay.
This program used as input a series of punched cards with six points
recorded on each card. Thus a computer program was necessary for the
conversion of the cards produced by the Flexowriter operation. In
February 1964, a simpler program became available using a single-
point-per-card format as the input. To eliminate the need for a
transparent overlay, the base map of Southern Ontario was preprinted
on continuous forms, and in March 1964, a map was produced with both
the symbols and the legend added by the tabulator (Fig. 4). This type
of map was similar to that used in the Atlas of the British Flora except
that the symbols available on the tabulator were not considered suitable
for reproduction in a publication.

Meanwhile, three other mapping tests were run in late 1963 and
early 1964 on two types of plotters. The first type of machine (Gerber
X-Y Plotter, Model GP-30-D and Hewlett-Packard Plotter, Dymec 2031)
used symbols on a printing head positioned by digital input signals which
were encoded on paper tape (Fig. 5). The second machine (CalComp
Plotter, Model 653) actually drew the symbols by means of a moving pen
and the input was by means of a deck of 80-column punched cards (Fig. 6).

These first experiments confirmed the fact that a tabulator is ideal
for distribution data which can be represented in relation to a regular

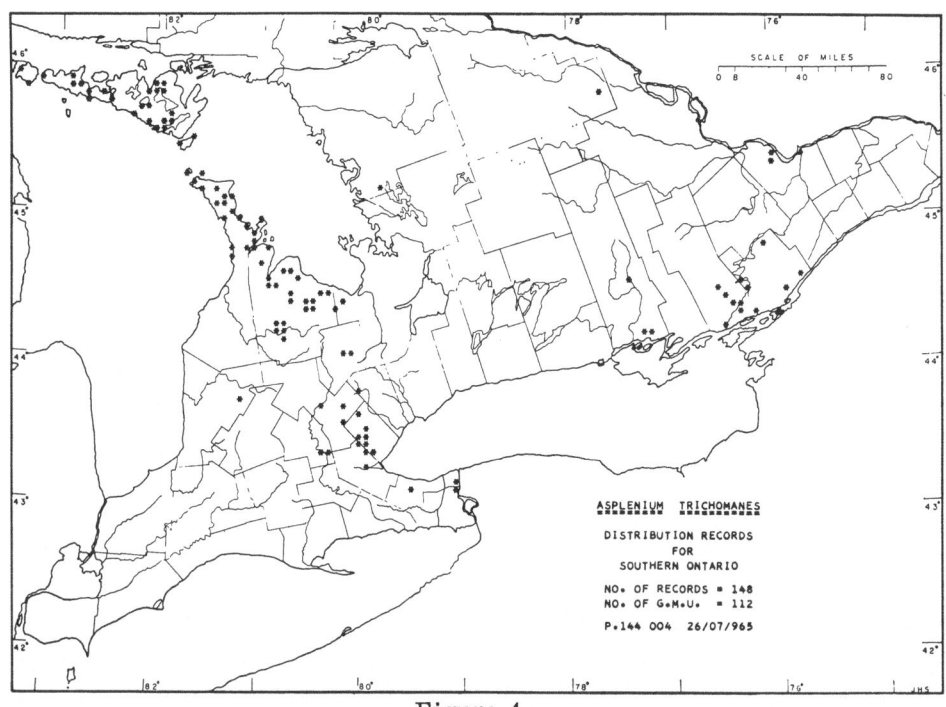

Figure 4

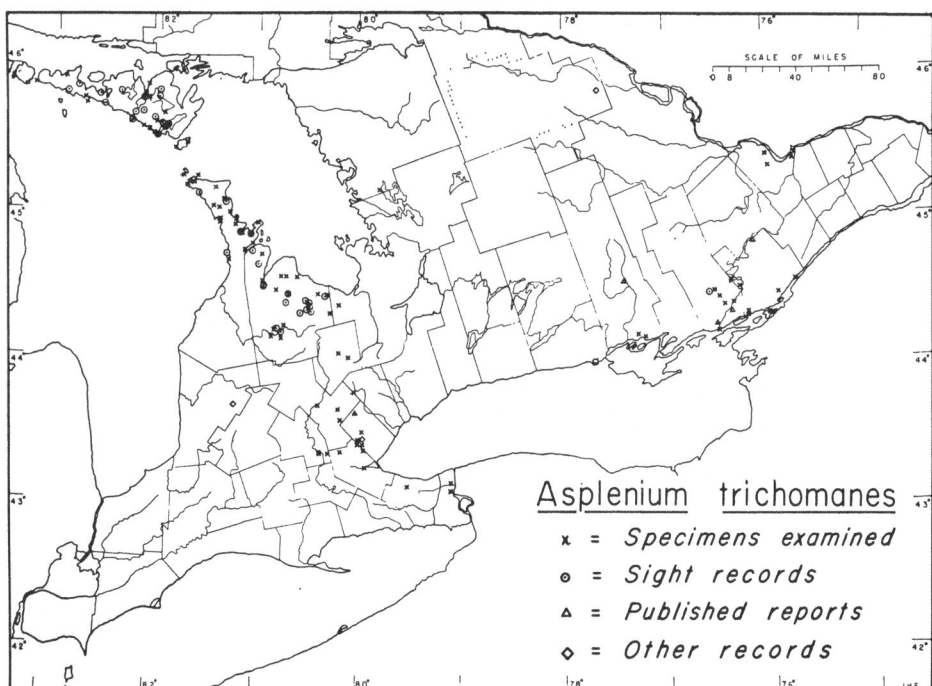

Figure 5

rectilinear grid and that a digital plotter (symbol-printing or line-drawing type) is required for making "dot maps" on which exact locations are to be shown.

In connection with the use of a tabulator for making distribution maps, it should be pointed out that the grid should correspond to a published topographic grid system in use for the area concerned. If the area is small, the Universal Transverse Mercator (U. T. M.) Grid may be used. This was tested with data for botanical collections on Manitoulin Island, Lake Huron. The base map of Manitoulin Island was prepared at such a scale that a grid of one inch squares coincided exactly with the U. T. M. grid overprinted on the corresponding maps in the National Topographic Survey series. It was then possible to take the U. T. M. grid readings of the "X" and "Y" co-ordinates ("Northing" and "Easting") directly from the topographic sheets at the scale of 1:50,000. When a gridded map is not available, it is necessary to place an arbitrary grid over the base map to be used for the plotting of records and to determine individually the co-ordinates of each point to be plotted. This results in a purely artificial gazetteer for that particular map which makes machine-mapping possible. However, the checking of locations is a very time-consuming job and the gazetteer cannot be used for any other map.

The final phase in testing different methods for mapping involved the digitising of the required base map, i.e. converting the map outline to a series of points which could then be presented to a line-plotter which draws the map. Fortunately, digitising equipment exists to mechanise the process of recording the points which define the map outline. The program developed includes a subroutine which converts the co-ordinates (latitude and longitude) of each record into the corresponding print position on the base map. The complete program will then instruct the plotter to draw the base map, plot all the records, using a variety of symbols, and add a specified legend. The current program called ONTMAP will produce titled maps for the distribution of a given species in Southern Ontario using a CalComp plotter (Figs. 7, 8, 9). The data are prepared on 80-column punched cards and an intermediate computer program examines the data and generates a magnetic tape which controls the plotter. When the number of maps to be plotted becomes large, it is probably more economical to have the base map preprinted on the forms so that the plotter adds the symbols and legend but does not have to draw the outline as well. Meanwhile, keeping the whole program intact allows for the plotting of the map at various

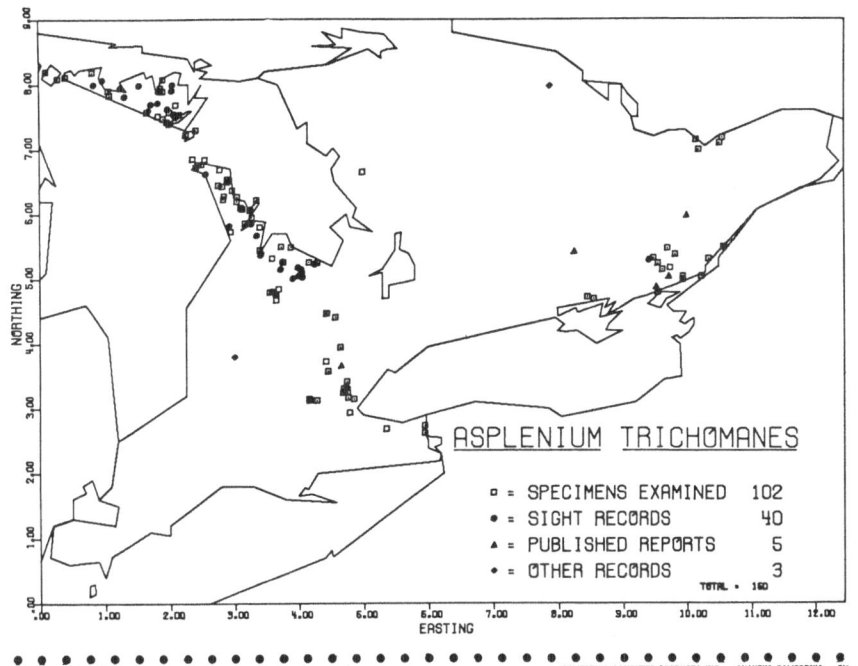

Figure 6

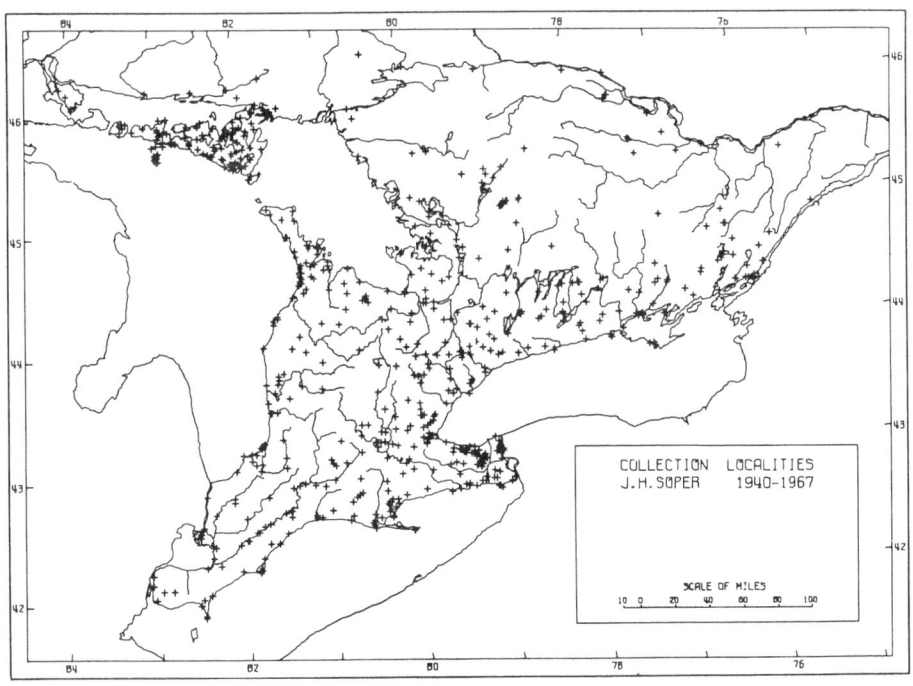

Figure 7

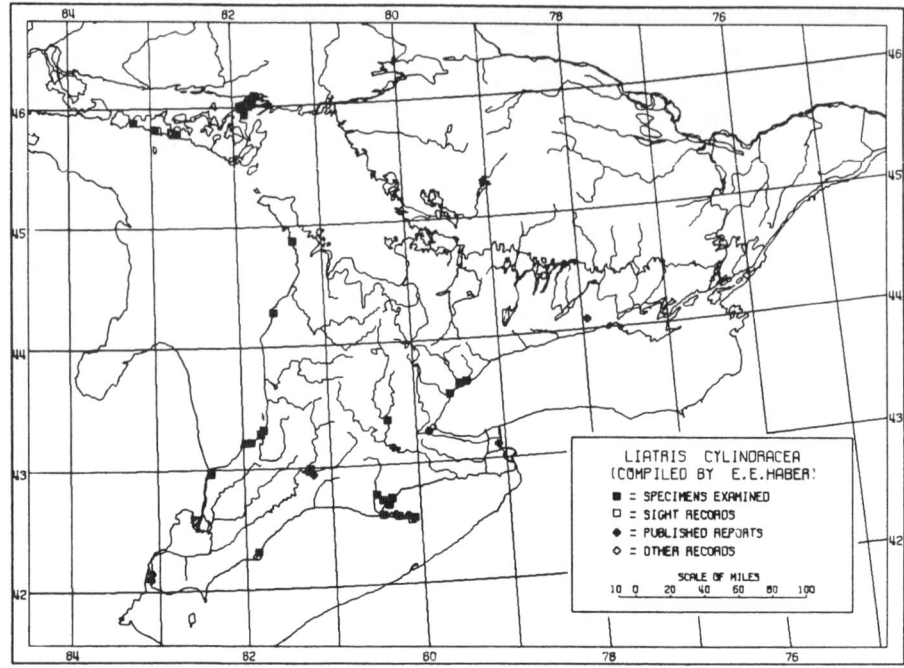

Figure 8

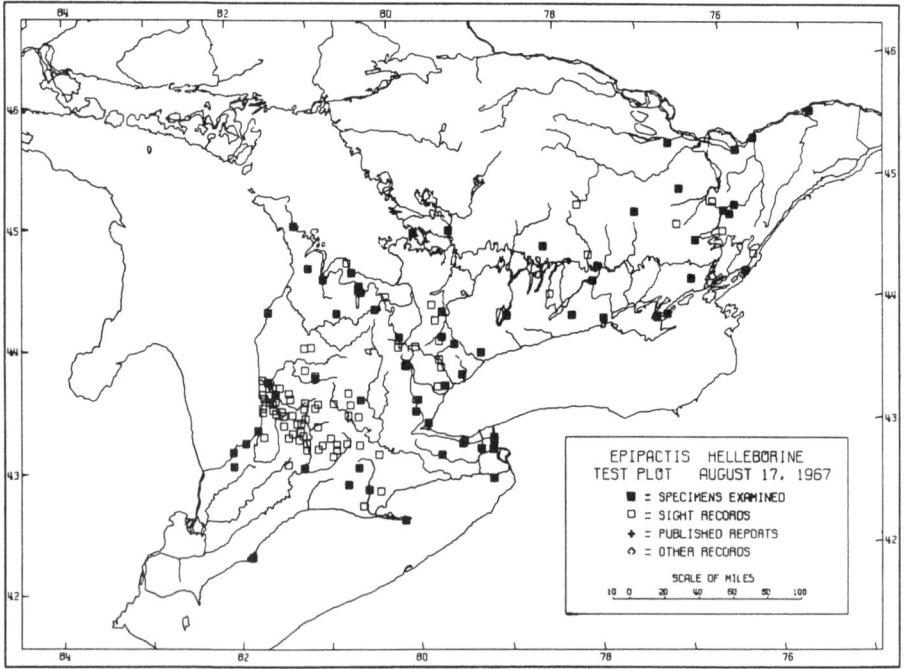

Figure 9

scales and for the addition of other modifications (Fig. 10).

2.   National Museum of Natural Sciences, Ottawa. At the National
Herbarium of Canada (CAN), the system developed at the University
of Toronto was set up in 1968 using the Friden Flexowriter and I.B.M.
card punch (Model 026) with mapping input tapes prepared on the
Government Computer Services Bureau's computer (I.B.M. 360) and
maps plotted on a 30-inch CalComp plotter in the same Bureau. On the
advice of a systems analyst, the data-capture via a Flexowriter and
card punch was discontinued in 1970 and an Alphatext terminal was
installed. This terminal consisted of a typewriter console (modified
I.B.M. Selectric typewriter) connected by telephone to a computer
located in Ottawa. On the keyboard of the terminal an operator
entered specimen records in a fixed format and the information was
received by the computer and held on magnetic disk storage until all
corrections had been made at the terminal. If necessary the
complete record could be printed out at the terminal, but when many
labels were required (some with several duplicates), the records
were transferred to a magnetic tape and printed later on special forms
using a high-speed line-printer.

Although the Alphatext system was ideal for text-editing and
therefore made correction of errors in label data quite simple, it did
not prove to be suitable either for data capture or for label production.
The preparation of additional programs to extract data for mapping or
to retrieve information in response to queries was therefore postponed.
The rather rigid input format and the lack of a simple method of dealing
with repetitive common data were distinct disadvantages in the system
and there were too many steps required between input and the final
printing of labels. More important was the fact that such an on-line
system if used on a full five-day-week (9 a.m. to 5 p.m.) basis proved
to be very expensive and the costs would soon have exceeded the budget
available for the operation of the pilot project.

Data-Capture. Distribution records accumulated during the use of the
Flexowriter and the Alphatext terminal have been converted to a common
file and stored on magnetic tape until they are required. The pilot project
dealt only with the vascular plants of Ontario and will be replaced by a
three-part system called Botanical Information Storage and Retrieval
(BISR) capable of dealing with phytogeographic data for the whole of
Canada and with records of all groups of plants.

The input device selected for BISR is a so-called "intelligent terminal" which can be operated off-line and is connected to a printer on which labels, catalogue records or other data in specified formats can be printed. Separate programs will extract records from a data bank and generate magnetic tapes for mapping or prepare lists or reports in response to specific queries. The machine is a Delta 5000 and has a C.R.T. display screen mounted above the keyboard of the console. Selected headings can be set up on the screen and common or fixed data can be held as successive specimen records are input individually. The operator can check the individual record after input and make corrections at any time during the input process. A small magnetic memory (2K) is associated with the device. As a record is being input on the keyboard it also appears in full on the screen. The complete record can be transferred immediately to magnetic tape storage on a cassette but it can also be printed on the printer which is attached to the console. Completed cassettes are subsequently transported physically to a computer centre and the data transferred to permanent storage in a data bank. An "update program" merges new records with those in the data bank and prepares a revised or updated file on magnetic tape.

Mapping. Work has also begun on the digitising of a large-scale base map of Canada suitable for reproduction in an atlas. It also appears feasible to have a more detailed map of Canada made available in digitised form with a program by which the map for a single province or any polygonal-shaped portion could be selected for plotting distribution maps on a provincial or regional basis. A preliminary test-map has already been digitised (Fig. 11).

Basic Data. The development and testing of various input and plotting devices has revealed the fact that many specimens in our herbaria have incomplete or inaccurate data on their labels. Very few specimen labels carry the exact co-ordinates for the geographic position of the site where the specimen was collected. Many labels give vague or incomplete descriptions of the locality. It is sometimes necessary to verify the latitude and longitude recorded on the label because typographical and map-reading errors are not uncommon. As more herbaria become involved in the use of E.D.P. for label production, map-making or information retrieval, the need for complete and accurate data on herbarium labels increases. Attention has recently been drawn to the need for some standardisation in the presentation of data on labels as a prerequisite to automation of certain herbarium procedures as well as for a general improvement in the quality of the information associated with herbarium specimens (Beschel & Soper, 1970).

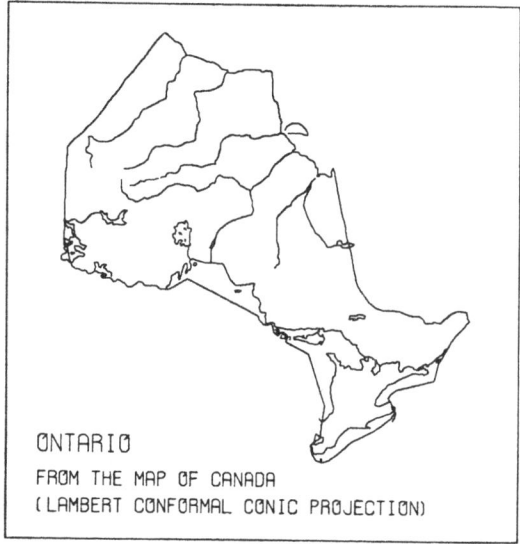

Figure 10

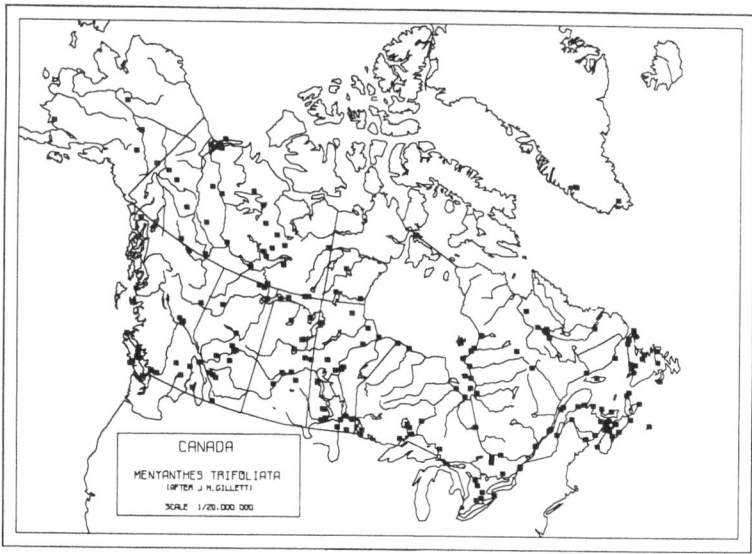

Figure 11

<u>Gazetteers</u>. As a by-product of checking data on collection localities in Ontario for mapping purposes, a gazetteer has been started which is proving very useful in machine-mapping. Every time the location of a collecting site is verified on the topographic map of the region, the latitude and longitude, locality name and description, date and collector's name are recorded in a file together with the reference to the map sheet used to locate the site. These entries are stored on magnetic tape and an updated print-out of the complete gazetteer can be obtained periodically. If each herbarium would build up a file of verified locality data relating to its principal collections and to collecting sites visited by members of its staff, a series of very useful gazetteers could be made available which would speed up considerably the preparation of distribution maps either by hand or by E. D. P. techniques. Such gazetteers would also be a source for the preparation of itineraries of collectors. A sample is appended (Table 8) which shows the format of the gazetteer being compiled for Ontario. In Canada there are published gazetteers available for individual provinces and territories. These can be used when the localities given on labels are named towns, villages or geographic features listed in the gazetteers. However, collections are more frequently made some distance away from a named locality and when the information is not precise it takes considerable time to locate the spot on an appropriate map. If the results of such searches were recorded and made available to plant geographers, it would mean considerable reduction in the duplication of effort which now occurs when the same specimen is consulted by more than one botanist. Regional gazetteers could be published or combined in one data bank in a central location with provision for access by all contributors and by others needing the information.

Table 8

```
MITCHELL                              ONT/Perth                 St. Marys 40 P/6 E
    43 28 N    081 12 W
    Fullarton & Logan Tps.
Mitchell #1                           ONT/Perth                 St. Marys 40 P/6 E
    43 29 N    081 13 W               G.R.Thaler                08/08/966
    Lot 20, conc. II, Logan Tp.  1 mi. N of Mitchell.
MITCHELL BAY                          ONT/Kent                  Chatham 40 J/8 W
    42 29 N    082 24 W
    Dover Tp. [settlement]
Mitchell Bay #1                       ONT/Kent                  Chatham 40 J/8 W
    42 26 N    082 25 W               J.K.Shields Loc. 26A,B,C. 12/08/950
    2.25-3.25 mi. SW of Mitchell Bay.  Dover Tp.  Lots 4,5,6, Conc. XI, XII.
Mitchell Bay #2                       ONT/Kent                  Chatham 40 J/8 W
    42 27 N    082 24 W               J.K.Shields Loc. 27       14/08/950
    1 mi. SSW Mitchell Bay, along dredge cut.  Dover Tp.  Lot 9, conc. XII.
    [also Loc. 24, 1.5 mi. SSW, Lots 7-8]
Mitchell Bay #3                       ONT/Kent                  Chatham 40 J/8 W
    42 28 N    082 22 W               J.H.Soper Loc. 54         11/08/941
    Dover Tp. Lot 15, conc. XII.
Mitchell Bay #4                       ONT/Kent                  Chatham 40 J/8 W
    42 28 N    082 24 W               J.K.Shields Loc. 31       22/08/950
    Road along dredge cut.  Dover Tp.  Lot 10, conc. XII.
Mitchell Bay #5                       ONT/Kent                  Chatham 40 J/8 W
    42 28 N    082 26 W
    Lake St. Clair [Bay]
Mitchell Bay #6                       ONT/Kent                  Chatham 40 J/8 W
    42 29 N    082 24 W               J.H.Soper Loc. 53         11/08/941
    Dover Tp.  Lot 14, conc. XIII.
Mitchell Bay #7                       ONT/Kent                  Chatham 40 J/8 W
    42 29 N    082 25 W               J.K.Shields Loc. 30       18/08/950
    Cul de Sac. 2.75 mi. NW of Mitchell Bay.  Dover Tp.  Lot 11, conc. XIV.
Mitchell Bay #8                       ONT/Kent                  Chatham 40 J/8 W
    42 30 N    082 24 W               J.H.Soper Loc. 52         11/08/941
    Dover Tp.  Lot 15, conc. XIV.
MOBERT                                ONT/Thunder Bay           Cedar Lake 42 C/12
    48 41 N    085 38 W
    Tp. 71.
MOIRA                                 ONT/Hastings              Tweed 31 C/6 W
    44 21 N    077 24 W
    Huntingdon Tp.
Moira #1                              ONT/Hastings              Tweed 31 C/6 W
    44 21 N    077 27 W               J.H.Soper Loc. 526        27/05/956
    Huntingdon Tp. Lot 4, conc. III. 3 mi. W of Moira.
MONCRIEFF                             ONT/Huron                 Seaforth 40 P/11 E
    43 39 N    081 09 W
    Grey Tp.
Moncrieff #1                          ONT/Huron                 Seaforth 40 P/11 E
    43 39 N    081 08 W               G.R.Thaler Loc. 52        16/08/965
    1 mi. N of Moncrieff.  Lot 28, conc. XIV, Grey Tp.
MONKTON                               ONT/Perth                 Seaforth 40 P/11 E
    43 35 N    081 05 W
    Logan & Elma Tps.
Monkton #1                            ONT/Perth                 Seaforth 40 P/11 E
    43 31 N    081 03 W               G.R.Thaler Loc. 159       17/08/966
    Lot 32, conc. XI, Ellice Tp.  5 mi. S of Monkton.
```

Conclusions

Computer-mapping is a worthwhile approach under certain conditions,
for example, if a large number of maps is required at one time, if a
map is to be updated repeatedly, or if a map is to be prepared at
different scales or with different combinations of data. The amount of
data to be plotted should be large and in a machine-readable format or
capable of being converted readily to such a format. It is also essential
that the data to be plotted carry geographic codes or location identifiers.

If a particular application satisfies some or all of these criteria, the
choice of equipment and program will be guided by the type of map
required, the accessibility of computer facilities and the funds available
to pay for programming services charges, and rental, lease or purchase
of the necessary machines.

When the area is small and a rectilinear geographic grid system
is available, a simple system using a tabulator with punched card,
paper tape or magnetic tape input is satisfactory to plot on preprinted
base maps. If special symbols are required, a line plotter may be
needed and if data are to be analysed before symbol selection can be
made, then a computer must be included in the system. When the map
outline is to be drawn together with the plotting of symbols, a line
plotter is required, usually with access to a computer to prepare the
input tape. It is apparent that the computer will be used more in the
future to analyse data, check for correlations between the distribution
of organisms and the factors of their environment and prepare
instructions for presenting spatial correlations on a map drawn by
automated cartographic means.

REFERENCES

ARGUS, G.W. & SHEARD, J.W. (1972). Two simple labeling and data retrieval systems for herbaria. Can. Journ. Bot. 50: 2197-2209

BESCHEL, R.E. & SOPER, J.H. (1970). The automation and standardization of certain herbarium procedures. Can. Journ. Bot. 48: 547-554.

BRITTON, D.M. & SOPER, J.H. (1966). The cytology and distribution of Dryopteris species in Ontario. Can. Journ. Bot. 44:63-78.

BROWN, C.E. (1964). A machine method for mapping insect survey records. The Forestry Chronicle 40: 445-449.

CADBURY, D.A., HAWKES, J.G. & READETT, R.C. (1971). A Computer-mapped Flora. A study of the County of Warwickshire, Academic Press, London.

CHARPIN, A. & MONTHOUX, O. (1971). L'emploi de l'ordinateur pour la cartographie floristique de la Haute-Savoie. Bull. Soc. bot. Fr. 118: 793-800.

GOMEZ-POMPA, A. & NEVLING, L.I. (1973). The use of electronic data processing methods in the flora of Vera Cruz program. Contr. Gray Herb. 203: 49-64.

HAWKES, J.G., KERSHAW, B.L. & READETT, R.C. (1968). Computer mapping of species distribution in a county flora. Watsonia 6: 350-364.

PERRING, F.H. (1963). Data-processing for the Atlas of the British Flora. Taxon, 12: 183-190.

PERRING, F.H. & WALTERS, S.M. (1962). Atlas of the British Flora. Thos. Nelson & Sons, London.

SOPER, J.H. (1964). Mapping the distribution of plants by machine. Can. Journ. Bot. 42: 1087-1100.

SOPER, J.H. (1966). Machine-plotting of phytogeographical data. Can. Geogr. 10: 15-26.

SOPER, J.H. (1969). The use of data processing methods in the
    herbarium. An. Inst. Biol. Univ. Na. Auton. Mexico, Ser. Bot.
    40: 105-116.

SOPER, J.H. & PERRING, F.H. (1967). Data processing in the
    herbarium and museum. Taxon, 16:13-19.

TAYLOR, D.R.F. (1971). Computer mapping: a tool for the 1970's.
    Rev. Géogr. Montr. 25: 381-389.

Discussion

Dr. A.V. Hall. I would like to make a comment on Dr. Soper's paper.
Difficulty has been encountered in verifying map references, and it has
been found an advantage to name grid squares   which can be reprinted
by the computer as a check.

Dr. F.H. Perring. In the work on the British Plant Records Scheme,
county names serve much the same purpose.

THE LIVING PLANT RECORD SYSTEM AT THE ROYAL BOTANIC GARDEN, EDINBURGH

J. Cullen

Royal Botanic Garden, Edinburgh, Scotland

Summary

The structure and operation of the recently computerised living plant record system in operation at Edinburgh is described, and its advantages compared to other kinds of record system are discussed.

So far, during this conference, we have heard about a number of E.D.P systems, either running now, or proposed, which encompass a wide variety of tasks. I have to deal with a system that, by comparison with much that we have heard, is rather small and simple, but with very defined aims. We have already heard something about record systems for living plants from Mr. Brown; I want to deal in greater detail with the system in operation at Edinburgh, which holds the records for the Royal Botanic Garden itself, and also for its associated gardens, the Younger Botanic Garden at Benmore, near Dunoon, and (ultimately) the Logan Botanic Garden at Portpatrick near Stranraer.

The aims of a garden record system are basically very simple, and relate to two functions: a) garden management in the broad sense, and b) the dissemination to the scientific and horticultural communities of information about the living plant resources of the garden. Both these functions are fulfilled by the answers to a rather simple set of questions: what taxa are in the garden?; where are they in the garden?; from where did they originate? The questions themselves may be simple, but the answers to them may be more complex; and different arrangements of these answers may serve different purposes. For instance, in terms of garden management, the horticultural staff actually involved in maintaining the garden require for easy reference an alphabetical listing of the taxa by genera and species; they also require lists of the plants to be found in the individual garden locations. For the scientific staff, a list of the garden contents arranged by families is of more importance, as is one arranged according to the various collectors who have contributed material to the garden. Both of these listings, as well as one by countries of origin, are important in terms of the dissemination of information to other scientists. The scientific staff may also require listings of various individual taxa, such as families or genera. These, then, are the important services which a garden record system has to provide. Before going on to describe how this is actually accomplished at Edinburgh, a comparison between a garden record system and one for herbaria will perhaps be useful.

The first major difference between the two types of record system is that of size. The size of a living plant collection is limited - by climate, size of area, soil, relief, number of staff available, etc., and, on the whole, is likely to be smaller, in terms of units accommodated, than most herbarium collections. The number of samples of any particular species is also likely to be smaller, and the ancillary information that goes with them - locality, habitat, date and collector - is, in general, simpler, and is usually compiled by someone in the garden, so that it is all in the same language and the same general form. This contrasts strongly with the situation in herbaria, which are usually much larger, contain many more collections per species, and do not usually have lists compiled of their contents - herbarium labels may be in many languages and different forms, and may require a good deal of interpretation.

Secondly, in a living collection, plants are, paradoxically, dying: some have a limited life span, such as annuals or monocarps, and even the longer-lived plants eventually die. In contrast, herbarium specimens may potentially last for ever, apart from occasional accidents. Thus, in a garden record, as well as new accessions entering the system, older ones are being cancelled from it.

Thirdly, there is the problem of a plant's location; where it is in the garden is an important part of the record which has no parallel in most herbaria. Other matters, such as identification and re-identification, assignment of species to genera and genera to families, etc., are common to both herbaria and gardens, and must find their place in the record systems of both.

A final, more general difference is that, to some extent at any rate, a well-run herbarium is self-indexing: the specimens are arranged in systematic order, often with a geographical break-down superimposed. This is not the case in a well-run garden, where specimens are planted in arrangements which do not necessarily correlate with systematics or geography, but may be based on aesthetic or cultural considerations. Hence, a record system is of more immediate importance in a garden than in an herbarium, because there are no one or two obvious places to look for the specimens of an individual species: they may be anywhere within the confines of the garden.

An adequate garden record, therefore, designed to facilitate the best and fullest use of the plant resources in the garden, will have the following contents: records of the plant's name, as far as known, its origin, its family, its garden location, and other additional items of information which will vary in individual cases. Of the four features specified, the first, the plant's name, is provided by the taxonomic system, and is needed for reference by anyone using the record system (some, at least, of whom will have had no scientific training) and must be recorded whole. The third and fourth, the plant's family and garden location, can be easily coded. The second, however, the plant's origin, is more difficult; it may contain narrative material on the original location and habitat of the plant, its collector and his reference number, and details of intermediaries between the original collector and the plant's current situation. Some of this information, such as collector and collector's number, can be easily coded for simple recording. But provenance, habitat and donor are more intractable. Because of this, each garden record is usually given a number (usually when the plant first arrives at the garden) which provides a key for this type of information. Such a number is given to all plant material when it arrives at Edinburgh, and is known as the accession number. Its first two digits are those of the year of accession, followed by four other digits which are given in serial order: thus 730001 is the first accession recorded in 1973. Since 1958 each individual accession has been given such a number; before 1958 each batch of material was given an accession number - for example, 570547 represented a mixed collection of bulbs, containing various species of Ornithogalum, Allium, Scilla and Gagea, presented to the garden by Dr. P. H. Davis after one of his expeditions to Turkey.

The accession records are kept in order of receipt in accession registers; and these form the first stage of the record system. The accession registers at Edinburgh go back to 1893, and are bound into books as suitable quantities are completed. This part of the system has not been computerised.

The accession stage of the record system is much larger than the other stages in terms of number of units recorded. There are several reasons for this, the main one being that many of the plants accessed do not get any further: seeds may not germinate, or seedlings may damp off, cuttings may not root, or plants may not become established in the garden. The accession record is thus, to some extent, a record of what might have been rather than a record of what is.

For plants that survive their earliest stages in the garden, a further record is necessary. At what point in the plant's history in the garden this further record should be given, is questionable. The basic principle at Edinburgh has been to provide this record when the plant is moved from the propagation department to one of the other garden departments, with a permanent location in the garden display. This may be changed in the future, however, because certain plants are retained in the propagation department for long periods, particularly material of purely scientific interest. It is clearly necessary that such plants should be included in the records.

Before computerisation this second-stage record was kept on specially produced index cards. These were maintained in alphabetical order of generic and specific names, and carried the following items of information: name, family, country or area of origin, garden location, accession number, collector and his number (if any), a statement whether or not the identity of the plant had been checked, and the date of its planting in the location recorded. On the back of the card there was a place for remarks of a more general nature, which might include details of the plant's performance, or references to relevant literature. From these cards the display labels for the garden were produced, showing the plant's name, family, country of origin, accession number, collector and number.

This card record was begun in the 1930's by the then curator of the garden, David Wilkie; by 1969 it held about 27,000 cards; this very large number was partly due to the fact that dead cards were maintained in the system, though in a separate alphabetical suite. Such a large collection of cards meant that sorting of them into any order other than the original was out of the question: it was therefore impossible to obtain such useful items as a list of all the plants belonging to one family, or all the species planted in one bed, or the various other forms of classified information already discussed. The decision was therefore taken in 1969 to computerise the record system to provide greater flexibility in sorting, and to make possible the production of a list of the garden's contents for ultimate publication.

A preliminary survey of the records as they stood was made by Management Services (Organisation & Methods Division) of the Scottish branch of the Civil Service Department. In consultation with the records office in the garden, they put forward a series of proposals regarding the form of the computerised record system to be adopted, and designed new cards of a more appropriate type to replace the older ones; the computerised information was ultimately taken from these new cards. The whole range of computing activities was undertaken by Scottish Office Computer Services (SOCS), who wrote the necessary programs (using COBOL language) and ran them on their IBM 360 computer in Edinburgh.

Because of the need for the unique reference for each record, the accession number became of considerable importance, as this, with the use of the first four letters of the specific name as a check, was the most suitable feature for this use. A complication here was that, as mentioned above, pre-1958 records did not have unique accession numbers. The first task, therefore, was to find such pre-1958 records, and to adjust them by providing notional accession numbers so that each would have its unique reference. The sorting was carried out as the first part of the computing process.

The first stage for the garden's record office involved the transfer of all relevant information from the old-style cards to the new ones. This, of course, had to be done manually and was therefore rather slow, but did provide an opportunity for a visual scan of all the cards, which allowed certain errors to be corrected, and certain inconsistencies to be tidied up before the computing stage was reached. The information from these new cards was then transferred to punched cards, and ultimately to magnetic tape.

The record information as it goes to the computer consists of 10 fields. Of these, the accession number forms the unique reference point,

and the records are stored on the tape in serial order of accession
number. The fields used are:-

1)   Accession number (6 digits).

2)   Generic name in full, with a total usable space of 20 letters.

3)   Specific name in full, with a total usable space of 23 letters, plus an
     extra one for the hybridity sign (x) if necessary.

4)   Subspecific, varietal or cultivar name in full, with a total usable
     space of 22 letters.

5)   Botanical family, using a code based on the first four letters of the
     family name.

6)   The area of origin of the taxon, using a 3-digit code of both
     alphabetical and numerical signs, based on the geographical
     divisions used in the herbarium.

7)   The collector of the material, using a 3-letter code based on the
     first 3 letters of the collector's name, or 3 initials in the case of
     material collected on joint expeditions.

8)   The collector's reference number (6 digits).

9)   Garden location, using a specially designed 3-digit code, involving
     a letter for the garden area, and 2-digit numbers for the individual
     beds.

10) The year of planting in the locations mentioned (a 2-digit number).

The records in this form could be up to 114 digits long.

     Once all this information was on tape, a number of vetting
programmes were run to find errors and inconsistencies, which were
then referred back to the garden records office for correction. The first
print-outs of the corrected information were available in February 1973 -
about $3\frac{1}{2}$ years after the initiation of the project.

     Print-outs are available in several different classifications, and are
normally produced every June and December:

1)   By genus and species, alphabetically.

2)   By family, alphabetically.

3)   By garden location.

4)   By original collector.

5)   By country of origin.

6)   By family, alphabetically, excluding taxa of garden origin - this
      print-out is used as the basis for the catalogue of plants in the
      garden, and is produced only as required.

7)   By individual genera, families, locations, collectors, etc., as
      requested.

      Having set up the system, means are necessary for maintaining it
in an up-to-date condition. Without such maintenance the value of the
system rapidly diminishes. The print-outs, being produced every 6
months, give, at most, a 6 month backlog of error.

      Emendations to the record consist of the following types:

a)   New accessions, entering the record for the first time.

b)   Deaths.

c)   Re-identifications, name changes, re-allocations of species to
      genera or genera to families. Such changes usually follow the
      publication of monographic or revisionary work, or study by
      members of the scientific staff of the garden.

d)   Changes in garden location, which are notified weekly to the records
      office by the members of the horticultural staff responsible for the
      moves.

e)   Other miscellaneous changes, errors that have not been previously
      noted, etc.

At present, such alterations are running at the rate of about 1000 per
month. They are notified to the computing staff by means of cards,
filled in at the records office, of which there are 5 types, designed for

the individual purposes mentioned above. Most of these changes require
no comment, but one point to be noted is that deaths are cancelled from
the record entirely, and are not separately recorded as such.

The advantages of this system are manifest: the ability to sort the
records into various orders gives great flexibility and makes possible
the planning of a rational acquisition policy and the overall assessment
of the value of the collection as a plant resource for scientific and other
endeavours, as well as simplifying many aspects of garden management.

In reaching this stage there have, however, been a number of
difficulties, and some discussion of these seems justified. The process
of computerisation began in 1969, and the first tangible results appeared
in 1973. Of this four-year period very little was actually involved in
computer operations. Most of the time was spent in getting the raw
data into a suitable form for the computer. As a card record already
existed at the beginning it might have been thought that this would be
a simple and rapid operation, but experience showed that this was not
so. The preparation of the data can be a long job unless it can be
mechanised in some way.

A further disadvantage is the necessity for the writing out of all
the maintenance and amendment cards by hand, these then being punched
and finally converted to magnetic tape. This slow procedure might be
overcome eventually by access either to a card-punching machine or to
a terminal with direct access to the computer.

The total cancellation of 'dead' records is also a disadvantage.
The storing of these in a 'dead' file may help us in answering queries
from the general public, such as: will a particular species grow in
Edinburgh? ; or: are the collecting details of a particular species
(now dead) available?

The record cards themselves contain various extra items: details
of re-identification, references to literature or to illustrations, records
of propagation, etc., which, because they fall into no obvious pattern,
cannot be easily coded for recording in the computer. At present, these
are kept on the record cards in the garden, where they may be consulted
as necessary.

This, then, is the record system as it stands at the moment. Its
computerisation has allowed much greater use to be made of the
information than was possible before, with consequent advantages

in the planning of the maintenance and the future development of the garden. In the future we hope to be able to modify the system, to cut out some of the rather time-consuming operations that have to be done at the moment, and to extend the storage to include further useful items of information.

Acknowledgements

In setting up the computerised record system, the staffs of Management Services (O & M Division) of the Scottish branch of the Civil Service Department, and of Scottish Office Computer Services were deeply involved, and our thanks are due to them for continuing assistance and advice. Our thanks must also go to Mr. A. Evans, Assistant Curator, and Miss Dorothy Purves, Records Officer, to whom the main bulk of the work in the garden has fallen.

Discussion

Professor R. Santesson. I wish to ask Dr. Cullen how many staff are involved.

Dr. J. Cullen. Two, under the direction of an Assistant Curator.

Mr. J. Raynal. Was this a useful way to produce a seed list?

Dr. J. Cullen. Edinburgh no longer produces a seed list, but our system can be used to produced a catalogue of what is grown.

Dr. J. L. Cutbill. Has any attempt been made to cost the development of this system, including the time spent in preliminary discussion? Users are always interested in this aspect, as well as the disorganisation feared as a consequence of the introduction of E.D.P.

Dr. J. Cullen. It is difficult to assess the cost properly due to the accounting system used by the Scottish Office, but the Royal Botanic Garden, Edinburgh, was not charged for computer time. Accountants had estimated that something in the region of £7000 had been saved by doing away with manual recording.

<u>Mr. D.M. Henderson</u>. Only the Records Office was involved in the discussions and initial transfer of information to cards was carried out by garden staff during times when they were unable to work outdoors.

THE USE OF E. D. P. IN ZOOLOGICAL COLLECTIONS

D. B. Williams

British Museum, Natural History

London, England

The text of this paper is here given in abstract form only.

The reasons for employing E. D. P. in a zoological collection are the same as those for any other collection. They range from assistance in collection management to the compilation of data banks for use in research. The use of E. D. P. in collection management may be justified in two ways: the reduction of effort in initial collection of data, and the reduction of effort in later processing of data such as production of indices and similar aids to assist in the location of material. Methods of employing E. D. P. to effect these savings are reviewed.

The structure of the data is complex, and the requirements are discussed. The combination of the properties of the data and the processes to be performed on it require the availability of a computer system with certain well defined properties. These properties are discussed.

Finally conclusions are drawn as a result of operational experience, on the needs for an adequate data standard and its properties.

DISCUSSION, FRIDAY 5 OCTOBER, AT END OF MORNING

Dr. A. Rannestad. I wish to make a few important general points.

1. The use and issue of microfiche in large multiple sets, as suggested by Dr. Perring, is likely to be far too expensive.

2. There is an extraordinary multiplicity of data-recording systems and this should be simplified as far as possible. In particular the following points should be noted:-

   a. Retrieval combinations must be agreed.

   b. Systems adopted must as far as possible be compatible with one another.

   c. Essential descriptors must be agreed on, and additional descriptors should also be specified.

Professor C.H. Oppenheimer. I wish to stress the importance of standardisation as recommended by the Stockholm Conference on the Environment. Information stored will in the future be used for integrated studies of the total environment. Therefore the conclusions reached by this Conference will have far-reaching significance.

DISCUSSION, MORNING, SATURDAY, 6 OCTOBER

This, the final discussion period of the Meeting, focussed initially on the papers read on 5 October, and subsequently broadened to cover general aspects of the theme of the Meeting. It was conducted by Professor J. G. Hawkes, who invited discussion on the general themes of this Conference.

Professor C. Kalkman. I came as a sceptic, but I have been converted in only a minor way. I have been convinced by the results and am frightened by the amount of work involved. Administration can be facilitated, but this is not very important. It is only superficially true that herbarium labels are a valuable source of data; I am not impressed by ecological work based on labels, for instance and, what is also important, ecology is not a field in which herbaria must be active in the first place. The first task of the large herbaria is taxonomic research. The production of floras and monographs is not as good as it should be, because, among the other things, the workers are indulging in marginal research. I would rather see a monograph of one small family than a list of sheets of all species in a certain area, but I fear that the present interest in E. D. P. will encourage the reverse.

My fear is that E. D. P. in European Herbaria will lead European taxonomists further away from systematic and taxonomic research. Now this has been put as a very black picture in order to restore some of the balance in your minds about the pros and cons of the fascinating techniques we have been talking about.

To become more practical, I would like to make three points.

1.   If herbarium curators and herbarium staff feel they must introduce E. D. P. methods in their curatorial and administrative practice, it is their business, and they may even be right. This is however, not an international affair, but entirely domestic.

2.   I will make an exception for a World Type Register. This will be an extremely useful tool for every taxonomist. The investigation by the Smithsonian Institution will be very valuable here, and I believe the work involved will be entirely justified.

3.   It has been implied that putting the label data of every species
      in the world in the computer memory will be of great help to
      taxonomy, ecology, environmental biology, plant geography
      and other disciplines. I will deny no scientist the right to
      produce electronically, for example, a list of woody climbers
      growing in Nothofagus forest between 1,000 and 2,000 metres.
      These are the kind of lists we are heading for. I do deny
      however the right to produce such a list if the species names
      on it are not trustworthy; this is likely as no more than half of
      the specimens in herbaria have been subjected to revision in
      the last 50 years. I would exclude from an international data
      bank any specimen whose identity is open to doubt. I would
      always ask if there is any scientific purpose in putting a
      certain collection into a data bank; I am convinced that in the
      first place (maybe even the only place) plant geographical
      research will benefit from E.D.P. methods, but only if and
      when the material is correctly identified.

I am worried about the output in plane systematic research. I am
convinced it could be much better, and this is a good opportunity to place
my views before you.

Dr. J.T. Williams. I would like to disagree slightly with the last
speaker. One of the most fascinating things about this Conference has
been the interface between herbarium records and other types of
collections. In particular I refer to living plant collections and
resources. At these interfaces it is often most important to exchange
information. I cannot agree, therefore, that making information
available by E.D.P. methods is a domestic matter. In the field of
genetic resources this is an international matter. To the person
involved with the taxonomy and evolution of cultivated plants or collections
of genetic resources material, the information on the specimens collected
by Vavilov and his colleagues could be most important. In similar
situations the question of data processing and its effective utilisation is
no longer a domestic matter for individual herbaria.

Dr. D. B. Williams.   One thing we have got to get clearly in our mind is the difference between publication of floras, monographs etc. and data processing. I think there is confusion in many people's minds. We have been brought up (in the major institutions) to worship the very highest degree of accuracy we can achieve, and this for the very good reason that our floras and monographs, when published, have habitually stood as major works which have been referred to all over the world for perhaps the last 50 to 100 years or even more.

Now with data processing we are not talking about this. For one thing we are not making a publication at all in any real sense of the word, and secondly we have a medium that can be continuously updated. I would ask you to remember this. We are all very conscious that the materials we have available for data publication are, and are likely to be, imperfect in a very large number of ways. But we are not committing them to the printed page which will be for evermore entombed in libraries; we are committing them to something which is very much more like a living organism, and which has the potential of being continuously adapted and updated and brought into line with the latest results of studies all over the world.

This is something which will be a major task to accomplish, and I readily acknowledge this, but I think it is something we should recognise as a possibility and look forward to it with some anticipation and enthusiasm. One other point I would like to make is this. When we are talking about E D.P., we tend to wipe off all the curatorial expenses which are now being incurred, day by day, year by year, in our various institutions. We tend to say, here is a very large sum of money which will have to be spent, and at the same time to disregard the huge sums of money which we are already spending on curation. Now I think we should look at this very, very carefully, because it seems to me that there are real possibilities that we use some of the money, albeit maybe with some substantial additions from other sources, for doing our curating very much more effectively, both for the good of our own institutions and in the interests of making our materials and information available to colleagues all over the world through these kinds of processes.

Dr. J. W. Franks. I think that many of the speakers seem to have missed a good deal of the point of the discussions and perhaps even the whole purpose of this meeting. I come from a major collection which is very much under-staffed. Our major problem in helping people is finding the specimens for them or knowing where they are, and I think it is in

this region that, as far as the major collections are concerned, we
can get help from this sort of technique. At the moment there are
certain things you just cannot find if people want to have them to work
on. You have to say, "You'll have to come from Australia; you'll
have to spend a week going through our covers to look for the material".
This is extremely wasteful both in time and money and scientific effort.
I'm not suggesting that when you get your print-out of what is in a
collection that you will want to see everything that is there, that
you accept every scientific name and locality as gospel, but what it does
give you is a chance to see what might be there, and what might be of
use - something which at the moment you can have no possible idea of
in many collections.

Dr. S.W. Greene. We were being faced with the problem of a vast
and rapid influx of material that we couldn't handle adequately. So we
appointed a typist to type our labels. When we added a paper-tape
typewriter, the typist did the initial job we wanted but we got a good
deal more besides. Certainly we had to re-arrange the order in which
we carried out some curatorial practices but the benefits greatly
outweighed the initial inconvenience. Unless one has had some
experience of the benefits of data processing methodology, it seems to
be very difficult to give the necessary reassurance that one is going to
gain greatly by altering slightly the traditional way of doing things.

I would like to emphasise that the information which one is
retrieving is in fact working information. I don't think any reputable
taxonomist is going to publish a list of names from an unchecked
print-out - if he did his reputation would just collapse. So instead of
seeing a danger here I see an enormous advantage - in being able to
get at our information more quickly.

I would like to add one final point. There is another bonus which
could be very valuable. In a monographic revision, etc. traditionally
one cites all of the specimens examined with their full collecting details,
often ending up with pages and pages of citations. Such lists could be
substantially reduced in size and hence made more manageable if only
specimen reference numbers were given, and the remainder of the
details were readily accessible in a data bank. Therefore one not only
saves a great deal of time in checking and re-checking manuscript
typescripts but, very importantly, publication costs are reduced. This,
I think, is an advantage that people haven't perhaps altogether appreciated

Dr. R.K. Brummitt. It seems to me that there is an essential
difference to be made between curating a national herbarium which is
essentially a research organisation and a smaller project such as the
several that we've heard about this week where E.D.P. is already a
going concern and which seem to produce very useful results at a
possibly economic price. The projects that we have heard mentioned,
I think I'm right in saying, have worked with up to 30 - 40,000
specimens, I don't think more than that. They are all dealing with a
clearly defined field, usually a fairly restricted geographical area,
where the geography is fairly well-known to the people doing the input;
the collectors' names are possibly also fairly well-known - it's a clear-
cut project. When you come to a national herbarium you are dealing
on a totally different scale. The figure of 30 or 40,000 specimens
which have been put into these projects over a number of years is
possibly only a six-month input of current accessions in national
herbaria such as Kew or Leiden or other comparable institutions. The
scale of the project is so enormously different. I have done some
superficial mathematics to find out how the input of specimens is
going to be coped with. If we have a figure quoted for Kew of
something like 60,000 a year, this comes out at about 250 to 300 a day
or 40 per hour. This is of course 40 per hour for each working day.
How many machines are we going to need to simply put in current
accessions in a national herbarium? I would think not less than three.
In addition to this you have the problem, which I don't think has been
mentioned so far, of updating. At the moment, when you are in the
herbarium and you find a wrongly determined specimen you simply
put it into the right folder. Once the specimen has been recorded you
can't simply do that - it has to be reprocessed and the correct name
put on record. This might put the figure up from three machines to
four, but that is simply for running the current accessions without
starting on the backlog of 5,000,000 specimens. With seven machines
running this represents 80 years,work approximately before you are
going to get the backlog on record. I think a lot of major herbaria are
sympathetic to this idea. It would be nice to have the information
available, but the problem is simply the scale of the task that's
involved.

Professor J. Hawkes. Most new methods are inevitably worked out in a
small way to begin with by means of pilot schemes. At a later stage it
can then be seen how these can be developed into very much larger
projects.

__Dr. A. Rannestad.__ I would like to place a question mark against
Dr. Brummitt's reasoning. In a large collection like the one at Kew,
there must be a lot of plants which are misnamed or misplaced. To
check and correct these is an enormous task, and it seems
unbelievable that it can ever be completed, but you are working
systematically at it. Why? Because you want the records of the
collection to be as good and as accurate as possible. This is exactly
why you need an E.D.P. system, to make your records better and
more accessible. The benefit of E.D.P. in a small collection is, in
my view, marginal compared with the possible benefits in a large
collection.

Of course, you cannot get your data into the E.D.P. system
overnight. To get five million items into an E.D.P. system will take
time, but you have to start, even though the task is very large – as
you are now doing with your manual records. Incidentally, the work
with your manual records may very well be combined with your work
to computerise your records. The E.D.P. system can, of course, be
used as soon as you have a segment of your records in the system, but
the full benefit will only be realised when the major part of the records
is computerised. You will then be able to compare, sort and list data
for a very large number of items, which would not be possible with a
manual system.

However, the main question today is not if you need or should
introduce an E.D.P. system in your collection. I feel confident that
all major European botanical collections will introduce computerised
systems sooner or later. Thus the main question is: will the
introduction be orderly with a minimum number of common basic
descriptors and the basic structure of the systems be such that you
will have compatibility and thus be able to interchange data; or will
you finish with many incompatible systems. You should therefore
agree on the basic principles for such systems and descriptors, so that
if or when you introduce a computerised system it will be compatible
with other systems in use.

__Mr. R. Ross.__ There are two things I would like to say at this stage.
The first is a comment on what Dr. Brummitt has just said, and that is
this. In any major herbarium, as in any smaller and more specialised
one, the actual research projects that people are working on are
concerned with comparatively small numbers of specimens. In the
handling of the information about these and the preservation of it for

publications of monographs or regional floras, there are probably
considerable advantages to be gained from the use of E. D. P. for
manipulating the data. I think we all tend to have limitations and to
fail to realise fully the extent to which we are liable to get benefits
from an ability to manipulate as computers can. Dr. Brummitt's
views have got a lot of soundness behind them; nevertheless advantages
of introducing E. D. P. first of all in the areas where it will produce
immediate dividends are fairly obvious. If you are going to do that,
then you need to take thought for the long-term requirements of
continuing to build a data bank as other research projects are built into
it. The second comment I would like to make is one that I feel very
strongly about in connection with herbarium curation and which was
mentioned by Dr. Kalkman when he spoke about one of the main tasks
of the herbarium being the production of regional floras. At the
present time, even with the kind of arrangement that there is at Kew
and many other herbaria, it is still impossible in any major herbarium
to tell anybody how they can find the specimens from any particular
area in whose flora they are interested, other than by going through the
herbarium from end to end. They can go very quickly to see specimens
of those species and genera that they know occur in that area in which
they are interested, but they can't go to the material of those that they
don't know are represented in the area, and it is those that are probably the
most important for them to see. E. D. P. is our only hope in being able
to remedy that situation.

Professor A. Gomez-Pompa. The National Herbarium of Mexico is a
relatively small herbarium, approximately 200, 000 specimens, compared
to the national herbaria of most european countries. In spite of its
relatively small size it is large enough to be computerized as a whole.
The first attempts in this direction indicated that the effort required was
not worth it. Based on this experience we decided to computerize the
data only from those specimens that come from areas where we have
special research interests. We have developed a system for data
capture that can be used in all herbaria. The main problem is to set
priorities for data capture. We decided that high priority should be
given to type collections, recently studied materials, and groups that are
under investigation. We are certain that E. D. P. will help to get
specimens more widely studied as the knowledge of their existence will
be more openly disseminated.

Professor H. Demiriz. We are at the moment reorganising our small
but potentially important herbarium in Istanbul and are doing our best to

have a modern one. It would be useful for us and for all our European colleagues if we could be supported financially, maybe by NATO or UNESCO, in introducing computer methods.

Mrs. N. Goulandris. I represent the newest and probably the least extensive herbarium in Europe, and I am very much tempted by the results I've heard here, and by the opportunities that using E.D.P. offers to a new herbarium, such as ours. We do not have the problems that face a big herbarium, for we only have about 10,000 sheets. We can therefore institute new methods from the beginning.

Mr. J. Raynal. Firstly I would like to make one comment about the accuracy of information taken from herbarium labels. I agree that all of us wish to have the highest accuracy in data banks. However, a data bank is something which at first is of internal use. Information stored in the data bank is only afterwards to be distributed elsewhere. I would like to ask Dr. Kalkman if he actually worries very much about misidentified specimens in his herbarium. In most herbaria in Europe there are many such plants, but we are not worried very much about them. In a data bank they would not be troublesome. I would agree with Dr. Kalkman's general point of view. However, we are facing an enormous backlog of specimens accumulated over decades, and this distinguishes us from other sciences. However if we don't begin something now, what is going to be the situation in 20 or 30 years? We will be even more backward than we are now in the eyes of other scientists. We are all sure that the technical aspects must be improved over the next few years. We have to demonstrate that E.D.P. methods work in order to get more people who are specialists in E.D.P. methods and who will relieve the taxonomists from doing this work. One last point: I think E.D.P. methods are not only going to make our work faster, but also more efficient. But to open new fields, I am thinking for example, of biogeography; for example, who among the biogeographers here has attempted to deal with comparisons of areas of distribution on a really scientific basis? I am sure this work cannot be done by hand and I am sure that a computer can do this work easily.

Professor C. Kalkman. I would like to ask one question of Mr. Raynal. I'm not worried about misidentifications going into a data bank. A herbarium data bank digests only information about the specimens held in that herbarium and there is no harm in that. But if it's a record of a species at a particular altitude in a particular region with ecological notes and so on which may be used in ecological or pharmaceutical research, then I would object very much to unverified data being used.

Mr. J. Raynal. Perhaps I didn't make myself clear enough. There is of course a great difference between on the one hand information put into a data bank and on the other information extracted from the bank, which must be quite accurate.

Professor J.G. Hawkes. Even so, you can build in an acceptability index, or, in other words, a credibility index on the identifications - something which says, for instance, it is a specimen that's been mentioned in a monograph in the last ten years, or that has not received modern treatment and so on.

Professor J. Heslop-Harrison. I am of course a plant physiologist rather than a plant taxonomist. However, I'd like to draw an analogy from plant physiology. I am sure most of us have had at some time or other a physiology course and heard about the way enzymes operate. Now if a reaction is possible, activation energy may be needed to start it, after which, if it is exothermic, energy is released. The difference between the problems of small and large herbaria is like that between a reaction with practically no activation energy and one which requires a considerable push to get over the hump. In a case like that of Dr Greene, the most efficient thing to do is to start with an EDP system and the rewards flow in immediately. At the other end of the scale we have Kew, Paris, Leningrad and others with a grotesque curve - considerable energy input over a long period before eventually the dividends come. A time span of 30 years has been mentioned; we don't know yet what the total input over this period will actually be. All that can be said is that the sooner we select and apply the most effective methods, the sooner the pay-off will come. One of the reasons for holding this Conference was just to get some idea of how the balance might work out for the larger herbaria, and it is perfectly clear from what we have heard this morning in our open discussions and in private discussions held outside of this hall that most of those concerned with the larger collections are worried because they know that catching up with the back-log is going to mean the expenditure of much time and effort. But are we to assume that we are never going to surmount the initial hill so that we stay on our present level, or are we to acknowledge that the future is so long that whatever the initial costs they are bound to be more than balanced by the total return, if you like, over hundreds of years? I for one take the view that our institutions, however mortal we are, are potentially immortal, and from this viewpoint I say that we have an infinite time axis, and the integration of any positive return of an infinite axis is bound ultimately to exceed the initial input over a finite time period.

SUPPLEMENTARY NOTES ON TAXONOMIC INFORMATION IN RELATION TO E. D. P.

J. G. Hawkes

Birmingham University

England

The following notes indicate some of the sorts of information needed by writers of taxonomic monographs, as well as by biographers, plant collectors and explorers, and those engaged in the compilation of local, national and regional Floras. The information could be made readily available with E. D. P. methods but is almost impossible to obtain under present systems, except in very small herbaria or private collections, and even then only to a limited extent.

The notes are set out below in the form of questions, with the areas of interest given in parenthesis after each.

1. Which herbaria contain material of a given collector, i.e., where are his materials to be found, and in what quantities? (Biographical, historical and monographic studies).

2. Which herbaria contain duplicates of a given specimen? This is particularly relevant to the question of holotypes and isotypes. (Taxonomic research and herbarium curating).

3. Which families, genera, species, etc. have been collected from a specific country, province, area, county, etc? (Essential for the construction of national and regional floras).

4. Which species or groups in a herbarium are in greatest need of revision? A "reliability index" on the names given on labels would

make the extraction of this kind of information possible.
(Monographic work; herbarium curating).

5. What habitat reference can be obtained for a specific locality,
    based on all the specimens collected there? In monographic work
    the data are available readily only for the group being worked on.
    (To give a monographer or future collectors an idea of habitats and
    plant communities).

6. What species were collected before and after a certain collection
    by a given collector? Example: Smith 3784 is suspected to be a
    hybrid; Smith 3781, 3732, 3783, 3785, 3786 might contain putative
    parent material but may not yet be identified and may therefore not
    be found in the same folder; or Smith 3784 may have faulty or
    missing label data, apart from the collector's number. The other
    collections, spread through many different families and therefore
    not available to the monographer, may provide a clue on the
    locality of Smith 3784. (Invaluable to monographers).

7. Which herbaria contain holotypes, isotypes, etc. of species and
    infraspecific categories? (Herbarium curating; monographic
    studies).

8. What are the times of flowering in material from certain
    localities or regions? (To indicate appropriate time for proposed
    collecting expeditions planning to visit a specific locality or region).

9. What altitude data are available from different collections made in
    the same region? (To cross-check altitudes given by different
    collectors, so as to plan a collecting expedition efficiently;
    valuable for monographs).

Combinations of two or more of the above-mentioned types of
information can be made readily available under E.D.P. methods.
This is much more difficult to accomplish, and in some cases
impossible, under present recording and sorting systems.

If herbarium materials are stored in alphabetical sequence of
families, and within these of genera, and again within these of
species, the taxonomist would be able to find his way to a
particular family, genus or species much more easily than at
present, irrespective of which system of classification he is used
to. E.D.P. methods can be used to print out lists of current

herbarium holdings, arranged in any of the currently accepted systems (Engler, Bentham and Hooker, etc.) to give the information which is obtained at present only for one system by means of the positioning of the specimens on the shelves. Difficulties of arrangement within genera, when species are sequenced partly taxonomically and partly alphabetically, would also be overcome.

CONCLUDING REMARKS

J. Heslop-Harrison

Royal Botanic Gardens

Kew, Surrey, England

I would first like to thank you all for participating in this Conference. That so many busy people have been prepared to devote time to it surely shows how important the topic is for us all. I wish particularly to thank those who came to Kew, not as delegates from sister institutions seeking guidance for the developement of E. D. P. policy, but to impart to us information and advice arising from their own practical experience in setting up working systems. The Conference would have meant nothing without their vital contributions, and we are surely most grateful to them for their help.

I am particularly pleased to record our deep thanks to the NATO Eco-Sciences Panel for the financial aid which made this meeting possible at all, and which will continue in support of the Working Party now set up. As you all know, one of the jobs of the Organising Committee will be to arrange publication of the full proceedings, and again this will be financed by NATO.

The task of the Organising Committee responsible for arranging the programme has not been an easy one, but under the Chairmanship of Professor Hawkes it has been discharged with great efficiency. I am sure you will wish me to congratulate the Committee upon its work, and thank the members for the time they have unstintingly devoted to making the Meeting a success. And in these thanks I would like to include our own local Secretariat under Mr. E. Timbs, who have overcome various difficulties in getting documents ready for the sessions.

A concluding word about the world context of what we have been discussing. Several motions have referred to the need for financial support for developments in the E.D.P. field for plant collections, and the possible role of the United Nations agencies, including UNESCO, UNEP and MAB. I believe it is most important for those of us in the great Botanic Gardens and Herbaria to bear in mind that our services are in considerable demand now, and will become increasingly more so in the future, by many different types of users, as interest in the preservation and rehabilitation of the human environment increases. There is a distinct tendency for us to be introverted about our work, and to suppose that it is only to serve the needs of our own science. In the English idiom, we are too inclined to exist by taking in each other's washing. Whatever may have been the situation in the past, the context is quite different now. A great many organisations are interested in plants as one of the great world resources; and they are coming to understand that like all other resources the plant kingdom requires proper management and husbanding. Let me just remind you that organisations like FAO are concerned not just with cultivated plants but their wild relatives and the genetic resources of wild communities. Who knows better about these wild communities than the botanists who work in them and deal with the taxonomy of the organisms that make them up? The great plant-oriented institutions provide basic data for many organisations outside our walls - organisations concerned with agriculture, forestry, resource management and the greater problems of conservation. These bodies are beginning now to acquire substantial funds, on both national and international levels.

Those familiar with the recommendations of the United Nations meeting on the Human Environment at Stockholm last year will know that embedded in these are a number of proposals about improving the documentation relating to the environment and its living resources. The recommendations highlight the need for better information systems for wild flora and fauna and of course also for collections under human care. Now who is going to provide the input for this documentation - for building up the "inventories" referred to in the Stockholm documents - if not the taxonomists who are actually working on the material? Certainly not the administrators in UN offices in New York, Rome or Nairobi!

It is for us of the greatest importance to realise that our institutions can now be fitted in as essential links in a larger chain. We have the potentiality of making substantial contributions to the great human enterprises which I am optimist enough to believe will be undertaken before the end of the century in aid of our planet. The governments of nearly every person here present have agreed, in principle anyway, to the Stockholm recommendations. Such agreement must mean ultimately the allocation of new funds. In so far as plants are the basic producers of the biome, work on the plant kingdom must surely take a very high priority, and it will be the fault of botanists themselves if it does not. In so far as the aims of the new strategies include the preservation of plant diversity, the work of those whose study is that diversity should clearly have a proper share of support.

I conclude with an appeal. An appeal that we should not be introverted in our work, but should be prepared to go out and show the world at large what we are doing, and what its relevance is in the new era of environmental awareness. If we do this, I believe there is no fear that the funds we require - and indeed have called for in some of the motions passed this morning - will become available.

Thank you all again for your attention!

RESOLUTIONS PASSED BY THE CONFERENCE

The following resolutions were passed:-

This Conference agrees:-

1. That data-banks related to plant collections should have an identical minimal standard set of descriptors, in the first instance based upon herbarium label-data.

2. That a Working Party be set up to advise, in the first instance, European herbaria upon the sets of descriptors referred to in proposal 1.

3. That individual herbaria should co-operate with their national organisations (where available) to create widely acceptable standards.

    That national organisations should be encouraged to work through the International Committee on Museums (I. C. O. M.) towards an international standard for documentation.

4. To recommend to the Organising Committee of the Leningrad International Botanical Congress (1975) that the matter of E. D. P. in taxonomic collections in the widest sense, including living collections, seed-banks, etc, should be discussed at the Congress and that the desirability should be considered of setting up a permanent international Commission for the co-ordination of E. D. P. in such collections.

5. That the Working Party, in addition to carrying out its primary function, of advising upon the sets of descriptors, should also deliberate upon software and systems, and the possibility of establishing a pilot project in one or more European institutions.

6. That the Working Party shall be empowered to consider and advise national herbaria on the appropriate steps to be taken for forming an international type-register.

7. That to promote the start of data-banks on a national and international scale, it is necessary that funds are allocated by international organisations. This Conference recommends that an approach be made by the relevant bodies to the appropriate

United Nations agencies, in the first instance UNESCO and UNEP, in solicitation of support for work of this type.

8. That the Organising Committee responsible for the organisation of this E. D. P. Conference remains in being for a further period of time to make detailed arrangements for publication.

9. That the Organising Committee responsible for the organisation of this E. D. P. Conference remains in being for a further period of time to supervise the setting up of the Working Party. Suggestions will be entertained by the Organising Committee, who will set up the Working Party with due regard to geographical representation and technical qualifications.

Les délégués présents à cette réunion sont d'accord sur les points suivants:

1. Les banques de données concernant les collections botaniques devraient toutes offrir un ensemble minimal identique de catégories d'informations tirées en premier lieu des étiquettes de ces collections.

2. Une Commission de Travail devra être formée pour conseiller les herbiers européens sur la nature des catégories essentielles d'informations mentionnées ci-dessus.

3. Les herbiers devraient coopérer avec les organismes nationaux adéquats pour élaborer des normes susceptibles d'être largement adoptées.

   Les organismes nationaux devraient être conduits à travailler à la recherche de normes internationales de documentation par l'intermédiaire du Comité International des Musées (I. C. O. M).

4. Il est demandé au Comité organisateur du Congrès International de Botanique de Leningrad (1975) de bien vouloir introduire comme sujet d'étude le problème du traitement électronique de l'information dans les collections taxonomiques au sens large, y compris les collections vivantes, banques de semences, etc., et de bien vouloir considérer le besoin de former une Commission Internationale permanente pour la coordination du traitement électronique de l'information dans ces collections.

5. La Commission de Travail, en plus de sa fonction première consistant à définir les catégories d'information, devrait aussi se préoccuper des systèmes et programmes de traitement, et de la possibilité d'établir un projet-pilote dans une ou plusieurs institutions européennes.

6. La Commission de Travail aura le pouvoir d'étudier et de proposer aux Herbiers nationaux les phases successives menant à l'élaboration d'un Catalogue International des Types.

7. Pour permettre l'implantation de banques de données à une échelle
    nationale ou internationale, un financement par les organisations
    internationales s'avère nécessaire. Il est recommandé que des
    contacts soient pris par les autorités compétentes auprés des
    institutions appropriées des Nations Unies, au premier chef
    l'UNESCO et l'UNEP, pour solliciter un soutien en vue d'un travail
    internationel de cet ordre.

8. Le Comité Organisateur responsable de l'organisation de cette
    Réunion sur le Traitement Electronique de l'Information
    continuera à fonctionner pendant le temps nécessaire à la
    préparation de la publication de ses résultats.

9. Le Comité Organisateur de cette réunion continuera à fonctionner
    le temps de superviser l'éstablissement de la Commission de
    Travail. Les candidatures seront examinées par le Comité
    Organisateur, qui determinera la composition de la Commission
    de Travail en fonction tant de la représentation géographique des
    institutions que de la compétence technique des personnes
    intéressées.

Die folgenden Beschlüsse wurden verabschiedet:

Diese Konferenz stimmt zu:

1. dass Datenbanken für Pflanzensammlungen ein identisches Minimum genormter Sätze von Deskriptoren enthalten sollten, die in erster Linie auf den Daten der Herbaretiketten basieren,

2. dass eine Arbeitsgruppe geschaffen wird, um in erster Linie die europäischen Herbarien über die Sätze der Deskriptoren, die im Vorschlag 1. genannt sind, zu beraten,

3. das die einzelnen Herbarien mit ihren nationalen Organisationen zusammenarbeiten sollten, um Normen zu schaffen, die in weitem Umfange annehmbar sind,

   dass die nationalen Organisationen angeregt werden sollten, über das Internationale Komitee an Museen (I. C. O. M.) auf einen internationalen Standard für Dokumentation hinzuarbeiten,

4. dem Organisationskomitee des Leningrader Internationalen Botanischen Kongresses (1975) zu empfehlen, das die Sache der Elektronischen Datenverarbeitung im weitesten Sinne, einschliesslich lebender Sammlungen, Samenbanken u. s. w. , auf dem Kongress diskutiert und darüber befunden werden sollte, ob es wunschenswert sei, eine permanente internationale Kommission für die Koordination der Elektronischen Datenverarbeitung in solchen Sammlungen zu errichten,

5. dass die Arbeitsgruppe zusätzlich zur Durchführung ihrer primären Funktion, über die Sätze der Deskriptoren zu beraten, auch Rat geben sollte über Software und Systeme und über die Möglichkeit der Errichtung eines richtungsweisenden Projekts in einem oder mehreren europaischen Institutionen,

6. das die Arbeitsgruppe ermächtigt werden soll, die nationalen Herbarien bei geeigneten Schritten zur Aufstellung eines internationalen Typenregisters zu beraten und anzuweisen,

7. dass, um den Start der Datenbanken auf nationaler und internationaler Ebene zu fördern, es notwendig ist, dass Mittel von internationalen Organisationen zugewiesen werden – Diese Konferenz empliehlt, dass ein Versuch durch die zuständigen Körperschaften bei den entsprechenden Stellen der Organisation der Vereinigten Nationen, in erster Linie bei UNESCO und UNEP, unternommen werden soll, mit der Bitte um Unterstützung für Arbeiten dieser Art,

8. dass das Organisationskomitee, das für die Organisation dieser E.D.P.-Konferenz verantwortlich war, einen weiteren Zeitabschnitt bestehen bleibt, um nähere Vorkehrungen für die Publikation zu treffen,

9. dass das Organisationskomitee, das für die Organisation dieser E.D.P.-Konferenz verantwortlich war, einen weiteren Zeitabschnitt bestehen bleibt, um die Bildung der Arbeitsgruppe zu leiten – Das Organisationskomitee wird Vorschläge machen, wer bei gebührender Berücksichtigung der geographischen Vertretung und der technischen Qualifikation die Arbeitsgruppe zusammensetzt.

LIST OF CONTRIBUTORS

J. P. M. Brenan
   Royal Botanic Gardens, Kew, Surrey, England

R. A. Brown
   American Horticultural Society, Mount Vernon, U. S. A.

J. Cullen
   Royal Botanic Garden, Edinburgh, Scotland

J. L. Cutbill
   Sedgwick Museum, Cambridge University, England

A. Gomez-Pompa
   Universidad Nacional Autonomia de Mexico, Mexico

Mrs. D. M. Greene
   British Antarctic Survey, Birmingham, England

S. W. Greene
   British Antarctic Survey, Birmingham, England

A. V. Hall
   University of Cape Town, Cape Town, South Africa

J. G. Hawkes
   University of Birmingham, Birmingham, England

D. M. Henderson
   Royal Botanic Garden, Edinburgh, Scotland

J. Heslop-Harrison
   Royal Botanic Gardens, Kew, Surrey, England

J. F. Mello
    National Museum of Natural History, Washington, U.S.A.

F. H. Perring
    Biological Records Centre, Monks Wood Experimental Station,
    Huntingdon, England

A. Rannestad
    Scientific Affairs Division, NATO, Brussels, Belgium

D. J. Rogers
    University of Colorado, Boulder, Colorado, U.S.A.

R. Ross
    British Museum (Natural History), London, England

S. G. Shetler
    National Museum of Natural History, Washington, U.S.A.

J. H. Soper
    National Museum of Natural History, Ottawa, Canada

J. A. Toledo
    Universidad Nacional Autonomia de Mexico, Mexico

D. B. Williams
    British Museum (Natural History), London, England

LIST OF PARTICIPANTS

O. Almborn
University of Lund, Lund, Sweden

Miss R. C. R. Angel
Royal Botanic Gardens, Kew, Surrey, England

Miss Y. Aspland
Royal Botanic Gardens, Kew, Surrey, England

F. Bjorkback
Swedish Museum of Natural History, Stockholm, Sweden

C. Booth
Commonwealth Mycological Institute, Kew, Surrey, England

D. D. Brezhnev
N. I. Vavilov All-Union Institute of Plant Industry (VIR), Leningrad,
U.S.S.R.

R. K. Brummitt
Royal Botanic Gardens, Kew, Surrey, England

J. F. M. Cannon
British Museum (Natural History), London, England

D. F. Chamberlain
Royal Botanic Garden, Edinburgh, Scotland

G. C. S. Clarke
British Museum (Natural History), London, England

W. D. Clayton
Royal Botanic Gardens, Kew, Surrey, England

R. S. Cowan
    National Museum of Natural History, Washington, U.S.A.

T. W. Davies
    Ministry of Agriculture, Fisheries and Food, London, England

H. Demiriz
    University of Istanbul, Istanbul, Turkey

Miss E. Ebbels
    Royal Botanic Gardens, Kew, Surrey, England

J. Flanagan
    National Botanic Gardens, Glasnevin, Dublin, Ireland

J. W. Franks
    University of Manchester, Manchester, England

J. Gerloff
    Botanischer Garten und Botanischer Museum, Berlin, Germany

Mrs. N. Goulandris
    Goulandris Natural History Museum, Athens, Greece

P. S. Green
    Royal Botanic Gardens, Kew, Surrey, England

E. F. Greenwood
    City of Liverpool Museums, Liverpool, England

S. G. Harrison
    National Museum of Wales, Cardiff, Wales

B. J. Harwood
    Ministry of Agriculture Fisheries and Food, London, England

R. H. Hedley
    British Museum (Natural History), London, England

D. R. Hunt
    Royal Botanic Gardens, Kew, Surrey, England

C. Kalkman
    Rijksherbarium, Leiden, The Netherlands

F. Kollman
  Hebrew University, Jerusalem, Israel

I. T. K. Kukkonen
  University of Helsinki, Helsinki, Finland

J. C. Ledoux
  Museum Requien, Avignon, France

J. Leon
  Food and Agriculture Organisation, Rome, Italy

J. Lewis
  British Museum (Natural History), London, England

J. Lundquist
  Umea University, Vindeln, Sweden

N. Lundquist
  Institute of Systematic Botany, Uppsala, Sweden

F. G. N. Lupton
  Plant Breeding Institute, Cambridge, England

Miss E. Lusher
  Botanischer Garten und Institut fur Systematik der Universitat
  Zurich, Zurich, Switzerland

J. M. Mascherpa
  Conservatoire Botanique, Geneva, Switzerland

H. A. McAllister
  University of Liverpool Botanic Garden, Neston, England

J. McDaniel
  Central Computer Agency, London, England

R. D. Meikle
  Royal Botanic Gardens, Kew, Surrey, England

E. J. S. M. Mendes
  Centro de Botanica, Lisbon, Portugal

Mrs. I. Mendoza-Heuer
Botanischer Garten und Institut fur Systematik der Universitat
Zurich, Zurich, Switzerland

J. Mennema
Rijksherbarium, Leiden, The Netherlands

B. D. Morley
National Botanic Gardens, Glasnevin, Dublin, Ireland

C. H. Oppenheimer
University of Texas, Port Aransas, Texas, U.S.A.

J. A. R. de Paiva
Instituto Botanico, Coimbra, Portugal

G. Panigrahi
Botanical Survey of India, (at Royal Botanic Gardens, Kew, Surrey,
England)

R. J. Pankhurst
Cambridge University, Cambridge, England

B. H. Peterson
Botanical Museum, Goteborg, Sweden

H. A. Pinnock
Department of Education and Science, London, England

K. Rahn
Botanical Museum, Copenhagen, Denmark

J. Raynal
Museum National d'Histoire Naturelle, Paris, France

H. Riedl
Naturhistorisches Museum, Vienna, Austria

L. Ryvarden
University of Oslo, Oslo, Norway

R. Santesson
Umea University, Vindeln, Sweden

A. Schreiber
Botanische Staatssammlung, Munchen, Germany

Miss D. Scott
Biological Records Centre, Monks Wood Experimental Station,
Huntingdon, England

G. E. Shmaraev
N. I. Vavilov All-Union Institute of Plant Industry, Leningrad
U.S.S.R.

R. W. Smith
Q.D.A., London, England

S. Snogerup
Institute of Systematic Botany, Lund, Sweden

A. L. Stoffers
Rijksunivsiteit te Utrecht, Heidelberglaan, The Netherlands

D. C. Sutton
Commonwealth Mycological Institute, Kew, Surrey, England

Mrs. J. Szujko-Lacza
Natural History Museum, Novenytar, Hungary

P. Thompson
Royal Botanic Gardens, Kew, Surrey, England

B. R. Townsend
Ministry of Agriculture, Fisheries and Food, London, England

L. Vanhecke
Brussels Herbarium, Brussels, Belgium

K. Walther
Institute fur Allgemeine Botanik, Hamburg, Germany

B. Westcott
Plant Breeding Institute, Cambridge, England

T. Williams
University of Birmingham, Birmingham, England

B. de Winter
   Botanical Research Institute, Pretoria, South Africa

SUBJECT INDEX

213